# The 8051 Microcontroller

## Hardware, Software and Interfacing

**James W. Stewart**

*DeVry Technical Institute*
*Woodbridge, New Jersey*

D0139035

REGENTS/PRENTICE HALL
Englewood Cliffs, New Jersey 07632

**Library of Congress Cataloging-in-Publication Data**

Stewart, James W., [date]
   The 8051 microcontroller : hardware, software and interfacing /
James W. Stewart.
      p.    cm.
   Includes index.
   ISBN 0–13–584046-5 :
   1. Automatic control.  2. INTEL 8051 (Computer)  I. Title.
TJ223.M53S73   1993
004.164—dc20                           92–13060
                                        CIP

Acquisition editor: Holly Hodder
Editorial/production supervision: Fred Dahl, Inkwell
Prepress buyer: Ilene Levy
Manufacturing buyer: Edward O'Dougherty

© 1993 by Regents/Prentice Hall
A Division of Simon & Schuster
Englewood Cliffs, New Jersey 07632

Printed in the United States of America
10  9  8  7  6  5  4  3  2  1

Prentice-Hall International (UK) Limited, *London*
Prentice-Hall of Australia Pty. Limited, *Sydney*
Prentice-Hall Canada Inc., *Toronto*
Prentice-Hall Hispanoamericana, S.A., *Mexico*
Prentice-Hall of India Private Limited, *New Delhi*
Prentice-Hall of Japan, Inc., *Tokyo*
Simon & Schuster Asia Pte. Ltd., *Singapore*
Editora Prentice-Hall do Brasil, Ltda., *Rio de Janeiro*

*For Shannon and Susannah*

# Contents

Contents

# Preface

This text is designed to be used in several ways:

• As a stand-alone text for an 8051-based course in microcontrollers.

• As a supplement to texts based on "standard" chips, such as the 8085.

• As a self-study tool for those whose work requires familiarity with microcontrollers.

The strong emphasis of this book is on interfacing the 8051 to real world devices such as switches, displays, motors, A/D converters, and so forth. The text has extensive discussion of device characteristics as well as many typical interfaces, such as RS-232 and IEEE-488. Since understanding the details is important, there are many software examples. The code is written in fully commented 8051 assembler language.

Although not a strict prerequisite, it would be helpful if the student has some familiarity with personal computers and has taken introductory courses in digital devices. It is assumed the student is familiar with binary and hexadecimal numbers. Each chapter includes an introduction, a summary, and questions for review. A separate transparency manual is available for instructors.

The material in Chapter 5 and its associated appendix was written by Professor G. Thomas Huetter. It is based on the notes of a very successful system design course that Dr. Huetter has developed and taught at DeVry Technical Institute of Woodbridge, New Jersey.

I wish to thank the following people for their help on this project: Holly Hodder of Prentice-Hall, Fred Dahl of Inkwell Publishing Services, Tom Huetter of DeVry Woodbridge.

*Jim Stewart*
*Piscataway, New Jersey*

# Architecture of a Microcomputer

Upon completion of this chapter, you should be able to

1. Give a short history of computers
2. Understand and use the special terminology to describe microprocessors
3. Explain what a microprocessor is, what its main components are, and what they do
4. Explain what is involved in the concept of processor speed

## 1.1 INTRODUCTION

This chapter describes in general terms the various subsystems of a microcomputer and the ways they interrelate—that is, the architecture of the microcomputer. Terms are defined and general concepts stressed. A detailed examination of the 8051 architecture is left to later chapters. Many terms are introduced in this chapter so that the student can begin to ''learn the language''; the short definitions given are not intended to be exhaustive.

### 1.1.1 Some History

The history of the computer is relatively short. In the nineteenth century, Charles Babbage designed a machine to perform calculations according to a stored set of instructions. He was assisted by Ada, the countess of Lovelace, who was the world's first programmer.

Unfortunately, Babbage never completed a working model. In the twentieth century, the theory of mechanized computation was further developed by such people as John von Neumann and Alan Turing. An early digital computer built with relays was designed by Howard Aiken in 1944. Automatic machines were built by John Atanasoff during the late 1930s. The ENIAC, the most famous early machine, was built at the University of Pennsylvania during World War II. Completed in 1946, it used a room full of vacuum tubes and could work for only a few hours at a time.

The commercial use of computers had to await the coming of the transistor in the early 1950s. Integrated circuit technology of the 1960s dramatically lowered the cost and size of circuitry and brought about the minicomputer. The minicomputer started the move of computers out of large institutions and into the hands of individual users in laboratories and factories, although the cost was still too high for the average person to afford.

In 1971, something revolutionary happened in the world of computers. The first general-purpose device to contain all the basic parts of a processor on a single chip, the microprocessor, was born. It was the Intel 4004, a 4-bit central processing unit developed at Intel by Ted Hoff, Federico Faggin, and Stan Mazor. In the following year the 8-bit 8008 was introduced. Interestingly, Intel developed the 8008 as a special device for Datapoint Corporation, a terminal manufacturer. Although the corporation had specified its design, Datapoint subsequently decided not to use the device because of speed and cost considerations. Intel then tried to sell the 8008 as a general-purpose device and discovered that a need (and thus a potential market) did indeed exist for an integrated circuit processor. The 8080, an improved version, soon followed and found immediate application in a wide range of products. An enormous success, the 8080 was the processor used in the first "home computer," which was sold as a mail-order kit for $350 by a company called MITS.

Other semiconductor manufacturers soon introduced their own devices. As is typical of a new technology, many different designs were marketed, but not all were successful. The successes included the MOS Technologies 6502 (used in the Apple II), the Motorola 6800, and the Zilog Z80. The power of a computer was now available with only a handful of chips costing relatively few dollars.

By the 1980s, microprocessors had grown in speed and capability. Single-chip computers were available. Word size had grown from 8-bit to 16-bit to 32-bit. Sophisticated software was available at relatively low cost. Now it was possible for the average person to buy, for a few hundred dollars, computing power that would have cost a few million dollars only 30 years earlier. But the real revolution was the computerization of everything—from automobiles to zoological equipment—made possible by the microprocessor. Today, the combination of microprocessors with robotics, telecommunications, and artificial intelligence promises a future of exciting new applications. This book is an introduction to that future.

### 1.1.2 Parts of a Microcomputer

A *microcomputer* is a system made up of a microprocessor, memory, input/output, and an interconnecting bus structure (see Fig. 1.1) integrated into a single chip. The parts of a microcomputer are described in the following sections.

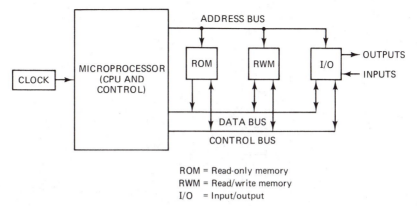

ROM = Read-only memory
RWM = Read/write memory
I/O = Input/output

**Figure 1.1** Block diagram of a microcomputer. (*Source:* Adapted from Kenneth Short, *Microprocessors and Programmed Logic,* © 1981, P. 12. Reprinted by permission of Prentice-Hall, Inc., Englewood Cliffs, N.J.)

### 1.1.3 Bits, Bytes, Words, Addresses

The basic unit of information in a computer is the *bit,* which stands for binary digit. A bit can have only the values 0 or 1 (low or high). A group of 8 bits is called a *byte.* A group of 4 bits, half a byte, is sometimes called a *nibble.* The *word length* of a processor is the size of the group of bits it is designed to use as a single unit or word. Thus, when speaking of an 8-bit, 16-bit, or 32-bit machine, we are referring to its word length.

A microprocessor requires various *registers* to hold groups of bits, some as big as the word size and some bigger. Some of the registers are also *counters* and can be incremented or decremented (add 1 or subtract 1), as well as have entire groups of bits moved into or out of them. One of the registers, the program counter (PC) (or instruction counter; IC), is used to hold binary numbers representing the *addresses* of instructions stored in memory. The size of the program counter, which is usually bigger than the word length, determines the memory space. The *memory space* is how many memory locations the processor can get to (address) directly and is equal to $2^N$, where $N$ is the number of bits in the PC register.

### EXAMPLE 1.1

A processor has an 8-bit word length and a 16-bit address length. How many words can it address?

**SOLUTION** $2^{16} = 65,536$. The 8085 can directly address 64K bytes of memory, where 1K is equal to 1024 (not 1000). Memory sizes are given in multiples of 1K.

## 1.2 THE CPU

The microprocessor itself contains a *CPU* (central processing unit), together with various counters, registers, and logic circuits required for the CPU to do its job. The central processor is the main component of a microcomputer; it's where the action takes place.

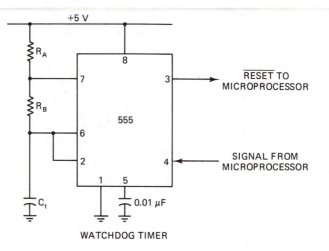

+5 V

$R_A$

$R_B$

555

8

7

6

2

1   5

3 → RESET TO MICROPROCESSOR

4 ← SIGNAL FROM MICROPROCESSOR

$C_t$

0.01 $\mu$F

WATCHDOG TIMER

### 1.2.1 Clock, Reset, Watchdog Timer

A CPU is a sequential state machine, and as such, it requires a *clock* to synchronize the internal transitions from state to state as it carries out its operations. The number of clock pulses required by the CPU to perform a basic operation is called a *machine cycle*. Also required is a *reset* input to put the CPU into a defined initial state when it is first powered up or to return the CPU to such a state from a severe error that locks up the processor.

Microprocessors built using NMOS semiconductor technology are dynamic devices. That is, they use the system clock input for internal refresh as well as timing and require a clock frequency higher than some specified minimum value to ensure proper operation. In contrast, the typical CMOS microprocessor is static and can be operated at an arbitrarily low clock rate. A CMOS microprocessor can be single-stepped by using a switch to generate clock pulses manually. The terms *static* and *dynamic* are explained further in Sec. 1.3.3.

Often a piece of equipment must operate with little or no human intervention. In such a case, if something happens to lock up the CPU, it is important that the system be able to recover by resetting the CPU itself. One technique is the so-called watchdog timer. A *watchdog timer* is a circuit that will wait a fixed amount of time for a signal from the CPU. If the signal arrives while the timer is waiting, it will restart the wait period. If the watchdog "times out" before it gets the signal, it will automatically reset the CPU. Such a circuit can be built with an integrated circuit such as the popular 555 timer.

### EXAMPLE 1.2

Use a 555 to design a watchdog timer. It should have a wait time of approximately 1 sec and produce a pulse of at least 0.1 sec.

SOLUTION   We will use the 555 in astable mode and assume that it causes a reset when its output (pin 3) goes low. The 555 output will stay high until capacitor Ct charges up. A periodic output signal from the processor will pulse the transistor to discharge Ct. We will arbitrarily pick Ct to be 1 μF. The 555 output is low during its discharge time (T2) given by T2 = 0.693 · Rb · Ct, so

$$0.1 = 0.693 \cdot Rb \cdot 1 \times 10^{-6}$$

$$Rb = 0.1/(0.693 \cdot 1 \times 10^{-6})$$

$$Rb = 144,300$$

We will make Rb 150K, which is a standard value close to 144,300. The charging time for Ct is given by T1 = 0.693 · (Ra + Rb) · Ct, so

$$1 = 0.693 \cdot (Ra + 150,000) \cdot 1 \times 10^{-6}$$

$$Ra = 1/(0.693 \cdot 1 \times 10^{-6}) - 150,000$$

$$Ra = 1,293,001$$

We will make Ra 1.5 Meg, which is a standard value close to 1,293,001.

## 1.2.2 Instructions, IR, MAR

*Instructions* are the commands, encoded as binary numbers and held in memory, that tell the microprocessor what to do. The main job of the CPU is to fetch (read) instructions and then execute them—the *fetch-execute* cycle. The address of the next instruction is held in the program counter. When the microprocessor is reset, the PC will contain some initial address, usually all zeros. As each instruction is fetched, the contents of the PC will be automatically increased (by an amount equal to the length of the instruction) so as to point to the next instruction in sequence. Some instructions (e.g., JUMP) change the contents of the PC when they are executed. Other instructions reference the memory to load or store data. Such instructions put addresses onto the address bus through the *memory address register* (MAR), which is similar to, but distinct from, the program counter.

Processors using the standard von Neumann architecture multiplex the PC and the MAR onto a common address bus and fetch both instructions and data over a common data bus. Processors using the so-called Harvard architecture have an address bus and a fetch bus for instructions and a separate address bus and read/write bus for data.

Once it is fetched, an instruction is held in the *instruction register* (IR) while the CPU decodes it. Some high-performance processors have an instruction *pipeline* (also called an instruction *queue*), which is a group of registers that allow a number of instructions to be *prefetched* and held while waiting to get into the IR. The prefetching is done while the CPU is executing the current instruction, and parts of the prefetched instructions may be executed simultaneously with the instruction in the IR in a technique called *overlap*. Getting an instruction from the pipeline is much faster than getting it from the memory, so the throughput of the CPU is increased by not having to wait for the next instruction. Similarly, some processors have an on-chip *cache* memory to hold frequently used data. Reading from cache is much faster than reading from external memory.

The general format of an instruction is an op-code followed by one or more operands. The *op-code* tells the CPU what operation to do; the *operands* specify on what piece of data the operation is performed. Instructions can be different lengths, usually multiples of a byte.

### 1.2.3 Accumulators, Scratchpads, Pointers

The typical CPU (but not all) contains one or more accumulators, equal in size to the word length. The *accumulator* is the main working register of the CPU. The result of an ALU operation (see Sec. 1.2.6) is left in the accumulator, where it may, in turn, become the operand of the next operation. Input and output usually flow through the accumulator.

A CPU also may have general-purpose word-length registers that can be used for temporary storage. These registers are sometimes called *scratchpads*. A CPU may also have one or more *pointer* registers (also called *index* registers) that are used to hold addresses. Special instructions can then get at memory locations by referring to the pointer instead of using the numeric address itself. Because the contents of a pointer can be incremented, decremented, or changed by the program, the pointer is a versatile feature. In some CPUs the scratchpad registers (sometimes in pairs) can be used as pointers.

### 1.2.4 The ALU

Some instructions move data around; other instructions perform calculations on the data. Calculations are done in a CPU by the *arithmetic and logic unit* (ALU). The ALU does adding and subtracting, as well as shifting, comparing, ANDing, and ORing.

### 1.2.5 Flags and Program Status

The CPU contains a special register, often called the *program status word* (PSW), that contains the *flag bits*. The flag bits are *set* or *cleared* (made 1 or 0), depending on results of CPU operations. Some instructions can "test" the flags, meaning that the execution of the instruction depends on whether a certain flag bit (or bits) is high or low. Some typical flags are sign, carry, zero, and overflow. Other flags, such as auxiliary carry and subtract, may also be present.

A flag bit usually refers to the state of the accumulator because most of the CPU work is done there. The *sign bit* is simply the *most significant bit* (MSB) of the accumulator after an ALU operation. When the content of the accumulator is being treated as a signed number, an MSB of 0 indicates positive and a 1 indicates negative. The *carry* flag can be thought of as an extra accumulator bit—for example, the ninth bit of an 8-bit accumulator. Any operation that causes a result exceeding the accumulator by 1 bit will set the carry; otherwise, the carry is clear.

*Overflow* occurs when the result of an arithmetic operation on signed numbers causes an erroneous carry into the sign bit. Because bits of the result are lost, overflow represents an error. The function of overflow should not be confused with that of the carry flag (see Chap. 4). The *zero* flag will be set by any operation that leaves all the bits in the accumulator equal to zero. The operation must be one using the ALU; just moving all zeros into the accumulator will not set the zero flag.

## 1.3 MEMORY

The set of instructions that tell the microcomputer what to do is called the *program*. The program, and perhaps tables of data, are stored in memory locations either on the microprocessor chip itself or in external memory chips. Memory devices are classified into two basic types: read-only memory (ROM) and read/write or random-access memory (RAM) (see Secs. 1.3.2 and 1.3.3). The 8051 contains both RAM and ROM.

### 1.3.1 Storage: Program, Data, Page, Segment

The total memory space can be divided into *program storage* and *data storage*. Such a division is usually just semantic because most processors treat memory as one continuous space. However, some processors (e.g., the 8051) have physically separate program and data storage with separate addresses.

Another division of memory space is into *pages* and *segments*. Typical sizes are 256 bytes for a page and 64K bytes for a segment. While often used simply as convenient descriptions, some processors actually handle memory in such fixed blocks.

### 1.3.2 ROM, EPROM, Volatility

As the name implies, read-only memory (ROM) devices can only be read by the microcomputer. To write into a ROM device requires a ROM programmer, which is separate from the microcomputer. Some ROM devices can be written to only once, whereas others can be erased and reprogrammed. A chip called a UVEPROM (or just EPROM) is an *erasable-programmable read-only memory* that is erased by exposing the chip to ultraviolet light in a special fixture.

ROM is nonvolatile, meaning that it does not lose its contents even when power is removed or it is disconnected from the system. Thus, ROM is often used to store the programs used by microcomputers. Programs so stored are referred to as *firmware*, which is derived from the terms *software* (the program) and *hardware* (the memory chip itself). In other words, firmware is software permanently embedded in hardware.

Another form of ROM is the *electrically alterable read-only memory* (EAROM). An EAROM is similar to an EPROM except that it can be programmed and erased under control of the microprocessor. Note that EPROMs and EAROMs can be erased and reprogrammed only a limited number of times before they fail.

### 1.3.3 RAM: Static, Dynamic, Battery Backup

The term *random access* was originally used to distinguish a memory device from one that was sequential access, such as magnetic tape. The term is now used to indicate that a memory device can be written to, as well as read, by a processor. The term *read/write* memory means the same as RAM.

Because a semiconductor RAM is essentially a large number of transistors on a chip, it is volatile. If power is removed, whatever was stored on the chip is lost. However, a RAM chip can be made nonvolatile by using *battery backup*. A battery is wired into the microprocessor system in such a way that during normal operation the RAM is powered by the system supply. But when power is lost, the RAM is automatically switched over to battery power, thus preserving its contents. The battery can be rechargeable, such as the nickel-cadmium battery, or it can be a primary cell type, such as the lithium. Some RAM chips are available with built-in batteries.

Another distinction is between static and dynamic RAM chips. *Static* memory devices store bits in bistable latch circuits (flip-flops) on the chip. In contrast, *dynamic* memory chips store bits in the form of stored charge on the gates of the integrated transistors. See Fig. 1.2. The charge is very small and tends to "bleed off," requiring that the bits be recharged periodically to maintain their data. The recharging occurs every few milliseconds when the processor accesses the memory chip during a special refresh cycle. Although the need for refresh cycles complicates the design of the memory subsystem, dynamic devices store more bits in less space and at lower cost than do static devices.

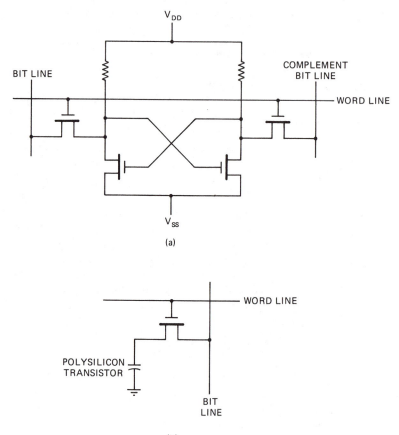

**Figure 1.2** Comparison of (a) static and (b) dynamic RAM cells.

Chapter 1: Architecture of a Microcomputer

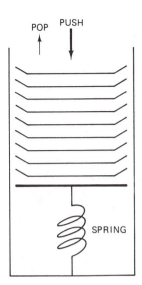

**Figure 1.3** Pictorial version of LIFO.

### 1.3.4 The Stack: Push, Pop, Call

Although RAM is used to hold programs and data, the CPU requires that a portion of RAM be set aside in the form of a *stack* for use by certain important operations. The stack gets its name from the stack of plates in a cafeteria. When a plate is taken off the top, another plate pops up to take its place. When new plates are added to the stack, the old plates are pushed down. The terms *push* and *pop* are carried over to memory stacks, which, in effect, work the same way. Stacks are described as being last-in, first-out, or *LIFO,* because the last thing pushed on will be the first thing popped off. See Fig. 1.3.

The stack is accessed sequentially by the CPU using a special address register called the *stack pointer* (SP), which points to the top of the stack. A push instruction will copy the contents of a register to the (hopefully) empty space at the top of the stack and the stack pointer will be advanced to point to the next location up the stack. A pop instruction will copy the top of the stack into a register and the SP will be adjusted to point to the next location down the stack. ("Up" and "down" are relative and depend on the processor.) The stack grows with every push and contracts with every pop. There must, eventually, be a pop for every push, or the stack will grow until it overflows—overwriting other parts of RAM and causing the system to crash. (The term *pull* is sometimes used for pop.)

The most important use of the stack is in a call to a subroutine. Briefly, a *subroutine* is a group of instructions written at one place in the program that can be executed from any other place. A *call* is an instruction that tells the CPU to go to the starting address of the subroutine and execute it. The last instruction in the subroutine is a *return,* which tells the CPU to go back to the main part of the program. The CPU knows the address of where it left the main program because a call instruction causes an automatic push of the return address onto the stack before branching to the subroutine. The return instruction causes an automatic pop of the return address off the stack and back into the program counter (assuming, of course, that the subroutine has not altered the stack). See Fig. 1.4.

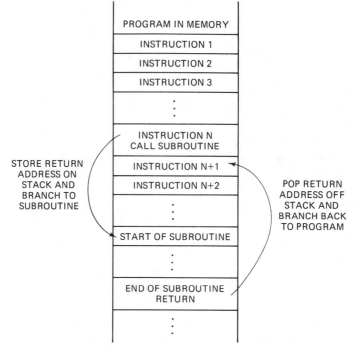

**Figure 1.4** Call to a subroutine.

### 1.3.5 Memory Speed and Devices

The speed of a memory chip is its *access time,* measured as the time between asking the chip for data and actually getting valid data from it. *Cycle time* refers to the time between successive accesses.

Access time can vary, depending on how the memory chip is asked. A typical ROM chip will have inputs called *chip select* (CS) and *output enable* (OE), as well as address and data lines. CS and OE are usually *active-low,* with CS being essentially an extra address pin and OE controlling output on the 3-state data pins. The term *3-state* (or *three-state*) refers to the three possible output states: high (1), low (0), and floating (high impedance or disconnected). Such devices are used to share a common bus. (*Note:* The term *tristate* means 3-state, but it is a registered trademark of National Semiconductor Corporation.) The CS signal is usually derived by decoding the higher address bits, and OE is driven by the CPU read control line (RD). See Fig. 1.5. Note that the CPU also generates a write control signal (WR) for writing to RAM devices.

In some memory devices, such as the 2764 EPROM (see Fig. 1.6), the chip-select function is called *chip enable* ($\overline{\text{CE}}$) and has an additional function of putting the chip into a low-power standby mode when not enabled, during which $\overline{\text{OE}}$ will be ignored. One way to access such a device is to give it a valid address, assert $\overline{\text{CE}}$, and then assert $\overline{\text{OE}}$ using

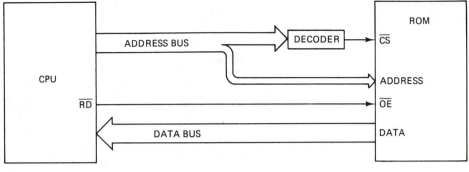

(a)

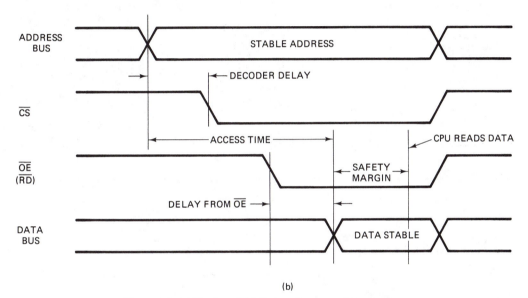

(b)

**Figure 1.5** (a) Block and (b) timing diagrams of access time.

$\overline{RD}$. An alternative is to ground the $\overline{CE}$ pin so that the chip is always enabled. To access the device, just give it a valid address and then assert $\overline{OE}$. Access time will be quicker at the expense of higher power consumption in the memory chip, an example of the speed-power trade-off common in digital systems.

    A typical RAM chip (see Fig. 1.7) does not have an $\overline{OE}$ pin, but it does have a read/write (R/W) control pin. Driving R/W to the read state causes the RAM to put data on the bus, similar to OE in a ROM. RAM chips typically have faster access times than ROM chips, which allows the R/W pin to be driven by the CPU RD line. See Fig. 1.8.

## Pin Names

| | |
|---|---|
| $A_0$–$A_{12}$ | Addresses |
| $\overline{CE}$ | Chip Enable |
| $\overline{OE}$ | Output Enable |
| $O_0$–$O_7$ | Outputs |
| $\overline{PGM}$ | Program |
| N.C. | No Connect |
| D.U. | Don't Use |

**2764A**
**P2764A**

| 27916 | 27513 | 27512 | 27256 | 27128A | 2732A | 2716 |
|---|---|---|---|---|---|---|
| $V_{PP}$ | D.U. | $A_{15}$ | $V_{PP}$ | $V_{PP}$ | | |
| $A_{12}$ | $A_{12}$ | $A_{12}$ | $A_{12}$ | $A_{12}$ | | |
| $A_7$ | $A_7$ | $A_7$ | $A_7$ | $A_7$ | $A_7$ | $A_7$ |
| $A_6$ | $A_6$ | $A_6$ | $A_6$ | $A_6$ | $A_6$ | $A_6$ |
| $A_5$ | $A_5$ | $A_5$ | $A_5$ | $A_5$ | $A_5$ | $A_5$ |
| $A_4$ | $A_4$ | $A_4$ | $A_4$ | $A_4$ | $A_4$ | $A_4$ |
| $A_3$ | $A_3$ | $A_3$ | $A_3$ | $A_3$ | $A_3$ | $A_3$ |
| $A_2$ | $A_2$ | $A_2$ | $A_2$ | $A_2$ | $A_2$ | $A_2$ |
| $A_1$ | $A_1$ | $A_1$ | $A_1$ | $A_1$ | $A_1$ | $A_1$ |
| $A_0$ | $A_0$ | $A_0$ | $A_0$ | $A_0$ | $A_0$ | $A_0$ |
| $O_0$ | $I/O_0$ | $O_0$ | $O_0$ | $O_0$ | $O_0$ | $O_0$ |
| $O_1$ | $I/O_1$ | $O_1$ | $O_1$ | $O_1$ | $O_1$ | $O_1$ |
| $O_2$ | $O_2$ | $O_2$ | $O_2$ | $O_2$ | $O_2$ | $O_2$ |
| Gnd | Gnd | Gnd | Gnd | Gnd | Gnd | Gnd |

Center device pinout (28-pin DIP):

| Pin | Signal | | Pin | Signal |
|---|---|---|---|---|
| 1 | $V_{PP}$ | | 28 | $V_{CC}$ |
| 2 | $A_{12}$ | | 27 | $\overline{PGM}$ |
| 3 | $A_7$ | | 26 | N.C. |
| 4 | $A_6$ | | 25 | $A_8$ |
| 5 | $A_5$ | | 24 | $A_9$ |
| 6 | $A_4$ | | 23 | $A_{11}$ |
| 7 | $A_3$ | | 22 | $\overline{OE}$ |
| 8 | $A_2$ | | 21 | $A_{10}$ |
| 9 | $A_1$ | | 20 | $\overline{CE}$ |
| 10 | $A_0$ | | 19 | $O_7$ |
| 11 | $O_0$ | | 18 | $O_6$ |
| 12 | $O_1$ | | 17 | $O_5$ |
| 13 | $O_2$ | | 16 | $O_4$ |
| 14 | GND | | 15 | $O_3$ |

| 2716 | 2732A | 27128A | 27256 | 27512 | 27513 | 27916 |
|---|---|---|---|---|---|---|
| | | $V_{CC}$ | $V_{CC}$ | $V_{CC}$ | $V_{CC}$ | $V_{CC}$ |
| | | $\overline{PGM}$ | $A_{14}$ | $A_{14}$ | $\overline{WE}$ | $\overline{PGM}/\overline{WE}$ |
| $V_{CC}$ | $V_{CC}$ | $A_{13}$ | $A_{13}$ | $A_{13}$ | $A_{13}$ | $A_{13}$ |
| $A_8$ | $A_8$ | $A_8$ | $A_8$ | $A_8$ | $A_8$ | $A_8$ |
| $A_9$ | $A_9$ | $A_9$ | $A_9$ | $A_9$ | $A_9$ | $A_9$ |
| $V_{PP}$ | $A_{11}$ | $A_{11}$ | $A_{11}$ | $A_{11}$ | $A_{11}$ | $A_{11}$ |
| $\overline{OE}$ | $\overline{OE}/V_{PP}$ | $\overline{OE}$ | $\overline{OE}$ | $\overline{OE}/V_{PP}$ | $\overline{OE}/V_{PP}$ | $\overline{OE}$ |
| $A_{10}$ | $A_{10}$ | $A_{10}$ | $A_{10}$ | $A_{10}$ | $A_{10}$ | $A_{10}$ |
| $\overline{CE}$ | $\overline{CE}$ | $\overline{CE}$ | $\overline{CE}$ | $\overline{CE}$ | $\overline{CE}$ | $\overline{CE}$ |
| $O_7$ | $O_7$ | $O_7$ | $O_7$ | $O_7$ | $O_7$ | $O_7$ |
| $O_6$ | $O_6$ | $O_6$ | $O_6$ | $O_6$ | $O_6$ | $O_6$ |
| $O_5$ | $O_5$ | $O_5$ | $O_5$ | $O_5$ | $O_5$ | $O_5$ |
| $O_4$ | $O_4$ | $O_4$ | $O_4$ | $O_4$ | $O_4$ | $O_4$ |
| $O_3$ | $O_3$ | $O_3$ | $O_3$ | $O_3$ | $O_3$ | $O_3$ |

230864–2

**NOTE:**
Intel "Universal Site"-Compatible EPROM pin configurations are shown in the blocks adjacent to the 2764A pins.

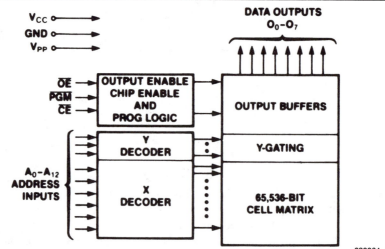

230864–1

**Figure 1.6** The 2764 EPROM. (*Source:* Reprinted by permission of Intel Corporation, Copyright © Intel Corporation 1988.)

## HM6264ASP-12, HM6264ASP-15, HM6264ASP-20, HM6264ALSP-12, HM6264ALSP-15, HM6264ALSP-20

**JANUARY, 1986**

### ⊚ HITACHI

8192-word x 8-bit High Speed Static CMOS RAM

**■ FEATURES**

- High Density 300 mil 28 pin Package
- Low Power Standby        Standby:  LP 0.01mW (typ.)
  Low Power Operation                    P 0.1mW (typ.)
- Fast access Time          Operating:  15mW/MHz (typ.)
- Single +5V Supply         120ns/150ns/200ns (max.)
- Completely Static Memory . . . No clock or Timing Strobe Required
- Equal Access and Cycle Time
- Common Data Inputs and Outputs, Three State Outputs
- Directly TTL Compatible: All Inputs and Outputs

(DP-28N)

**■ BLOCK DIAGRAM**

**■ PIN ARRANGEMENT**

(Top View)

**■ ABSOLUTE MAXIMUM RATINGS**

| Item | Symbol | Rating | Unit |
|---|---|---|---|
| Terminal Voltage * | $V_T$ | −0.5 ** to +7.0 | V |
| Power Dissipation | $P_T$ | 1.0 | W |
| Operating Temperature | $T_{opr}$ | 0 to +70 | °C |
| Storage Temperature | $T_{stg}$ | −55 to +125 | °C |
| Storage Temperature (Under Bias) | $T_{bias}$ | −10 to  +85 | °C |

* With respect to GND.    ** Pulse width 50ns: −3.0 V

**■ TRUTH TABLE**

| $\overline{WE}$ | $\overline{CS_1}$ | $CS_2$ | $\overline{OE}$ | Mode | I/O Pin | $V_{CC}$ Current | Note |
|---|---|---|---|---|---|---|---|
| X | H | X | X | Not Selected | High Z | $I_{SB}, I_{SB1}$ | |
| X | X | L | X | (Power Down) | High Z | $I_{SB}, I_{SB1}$ | |
| H | L | H | H | Output Disabled | High Z | $I_{CC}$ | |
| H | L | H | L | Read | Dout | $I_{CC}$ | |
| L | L | H | H | Write | Din | $I_{CC}$ | Write Cycle (1) |
| L | L | H | L | Write | Din | $I_{CC}$ | Write Cycle (2) |

X : H or L

**Figure 1.7** Typical RAM chip. (*Source:* Courtesy of Hitachi America, Ltd. This manual may, wholly or partially, be subject to change without notice.)

## 1.4 INPUT/OUTPUT (I/O)

To do something useful, a microcomputer must be able to exchange information with the real world—that is, hardware other than the memory and the CPU. Usually, a computer must *interface* with keyboards, displays, LEDs, switches, disk drives, and other such *I/O devices*.

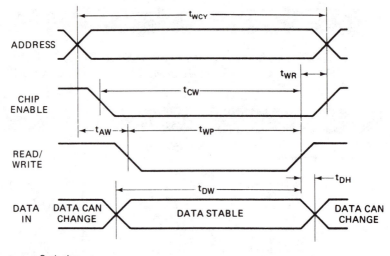

$t_{WCY}$ = Cycle time
$t_{CW}$ = Time chip select (chip enable) must be stable before
another transition on R/W
$t_{WR}$ = Time address must be stable before another R/W transition
$t_{AW}$ = Address to read/write delay (address setup)
$t_{WP}$ = Minimum read/write pulse width
$t_{DW}$ = Minimum data setup time
$t_{DH}$ = Data hold time after R/W change

**Figure 1.8** Write timing diagram for RAM chip. (*Source:* Adapted from Kenneth Short, *Microprocessors and Programmed Logic,* © 1981, P. 47. Reprinted by permission of Prentice-Hall, Inc., Englewood Cliffs, N.J.)

### 1.4.1 Ports and Memory Mapping

Microcomputers connect to I/O devices through *ports,* which are connections consisting of groups of parallel bit lines going into and out of the computer. A single-bit port is called a *serial* port. Ports often have addresses separate from memory addresses and special instructions to access them. Some systems use *memory-mapped I/O,* where the I/O devices are treated like memory locations, and all the instructions that access memory can then be used for input and output.

### 1.4.2 Latches, Drivers, Peripheral Devices

Often a separate chip containing *latches* and *drivers* is connected to the system bus to implement the I/O function of the microcomputer system. The data sent to a port address will go through the chip and into the real-world I/O device. See Fig. 1.9. Some I/O devices, such as a VART, are built into the 8051.

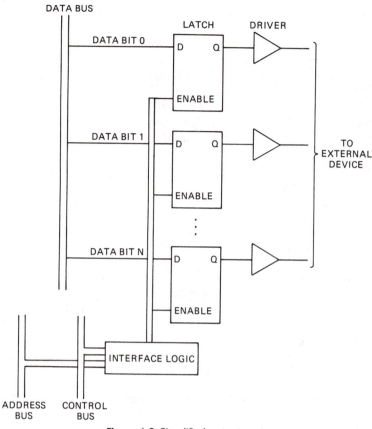

**Figure 1.9** Simplified output port.

### 1.4.3 Interrupts

Related to the idea of I/O is the idea of an *interrupt*. Microprocessors are often used to monitor and control external processes such as machine tools and communications links. In such systems, often called real-time systems, events will happen that require immediate attention from the processor. By activating the appropriate pin on the CPU or peripheral chip, an external device can interrupt whatever the CPU is doing and cause it to branch to an interrupt service routine. The process is similar to a subroutine call. Interrupts are examined further in later chapters.

## 1.5 SPEED

When comparing the speeds of two different microcomputer systems, it is tempting to look only at their CPU clock frequencies. For example, a 5-MHz processor should be "faster" than a 4-MHz CPU. But the speed we are really interested in is how fast the system does the application for which it was designed, not how fast the clock runs.

### 1.5.1 Throughput and Instructions

One aspect of speed is *throughput,* the number of instructions executed per second. The 4-MHz CPU may have a higher throughput than the 5-MHz if instructions in the 4-MHz unit require fewer machine cycles or if each machine cycle requires fewer clock pulses to complete.

Another factor is how well the CPU *instruction set* matches the program requirements. For example, if a program does a lot of numeric calculations (number crunching) but the CPU lacks specific MULTIPLY and DIVIDE instructions, then all the arithmetic will have to be done by repeated addition and subtraction. A CPU with the necessary instructions may do the calculations faster even though it has a slower clock. An alternative approach, known as a reduced instruction set computer (RISC), uses a small set of instructions that execute very rapidly. Conventional complex instruction set computers are sometimes referred to as CISCs.

### 1.5.2 Buses and Memory

The organization of the bus structure, at both the chip and package levels, has an effect on throughput. A 16-bit machine that uses an 8-bit bus will be slower than a 16-bit machine on a 16-bit bus, and may even be slower than an 8-bit machine on an 8-bit bus, depending on the architecture. The product of bus width and clock speed is sometimes called *bandwidth* and is a figure of merit when comparing speeds. Also, a fast CPU joined to a slow memory will have a lower throughput because it will be forced to execute wait states in any instruction that references memory.

### 1.5.3 Coprocessors and On-Board Peripherals

For systems used in applications requiring large amounts of specialized processing, throughput can often be greatly improved by using a special sort of peripheral called a *coprocessor.* As the name implies, a coprocessor chip is itself a microprocessor designed to do a specific job very fast. Whenever that job is called for in a program, the CPU gives it to the coprocessor, which does the job and gives the results back to the CPU. An example is the 8087 math coprocessor chip used with the 8086 CPU.

The term *peripheral* usually refers to a device, such as an A/D converter, that is separate from the microprocessor but communicates with it via the buses or through I/O ports. Thus the idea of an "on-board peripheral" may sound like a contradiction in terms. But thanks to advances in silicon processing, it is now possible to incorporate entire function blocks, such as A/D, onto the same chip as the CPU. Such function blocks can communicate with the CPU via the internal chip buses.

Thus we define *on-board peripherals* to be just such functional blocks. Many functions are available on various microcontrollers. Besides A/D there is also D/A, stepper-motor drive, digital signal processing, automobile functions, and many more.

### 1.5.4 Benchmarks

To really compare apples to apples in terms of speed, you want to compare how fast two different microprocessor systems will run a specific application. One way to do that is to write test programs called *benchmarks* that closely resemble the actual application. Each microcomputer will have the benchmark program written in its own instruction set, using its best features. The systems can then be run side by side and the times compared, giving a more valid measure of which is truly faster.

Standard benchmark programs have been developed for comparison purposes—for example, the Whetstone, which contains floating-point calculations, and the Dhrystone, which uses fixed point. However, the best benchmark is one that most closely matches your application.

## 1.6 SUMMARY

Microcomputers are systems consisting of these basic parts:

- The CPU, which does the actual processing
- The memory, which stores the program and the data
- The I/O, which allows access to the real world

The CPU was described as using the ALU as well as various registers and flags in order to fetch and execute the instructions. Regarding memory, we discussed RAM, ROM, and stack, as well as such terms as static, dynamic, and volatility. Under I/O we discussed ports and memory mapping. The concept of speed in a microcomputer was discussed, together with the speed-related issues of on-board peripherals and benchmarks.

### CHAPTER REVIEW

#### Questions

1. Who was the first computer programmer?
2. Who designed the first stored program machine?
3. When and where was the first electronic computer built?
4. When was the first microprocessor released?
5. Name the four basic parts of a microcomputer.
6. What is the difference between a byte and a word?
7. What do the following initials stand for: CPU, ALU, PC, IR?
8. Why does a CPU require a clock?
9. Why does a CPU require a reset?
10. What is held in the program counter?
11. What is the purpose of an instruction queue?
12. What part of an instruction tells the CPU what to do?
13. Describe the importance of the accumulator.

14. What is a pointer register?

15. What does the ALU do?

16. What is the PSW?

17. How are flag bits used?

18. How does the carry flag differ from the overflow flag?

19. What is the sign bit?

20. How is the zero bit set?

21. Explain the difference between RAM and ROM.

22. What is firmware?

23. What does nonvolatile mean?

24. Explain the difference between static and dynamic RAM.

25. Give an example of an application where battery backup would be needed.

26. Explain the operation of a memory stack.

27. What does LIFO stand for?

28. What is stack overflow and what might cause it?

29. What is the difference between port-addressed I/O and memory-mapped I/O?

30. Describe a possible application for a serial port.

31. What is a peripheral device?

32. Explain how a 4-MHz CPU can be faster than a 5-MHz CPU.

33. What is the advantage of using a coprocessor?

34. Explain the use of benchmark programs.

35. Why are interrupts necessary?

## Problems

1. How many locations can a CPU address directly if it has a 24-line address bus?

2. Design a watchdog timer using a 555 that will wait 500 msec before generating a 200-msec active-low reset pulse.

## Chapter 2

# The 8051 Single-Chip Microcontroller

Upon completion of this chapter, you should be able to

1. Understand and use the special terminology of a microcontroller
2. Describe the function of every pin on the 8051
3. Describe the 8051 memory organization
4. Describe the uses of the various special function registers
5. Describe the Boolean processor
6. Describe the 8051 instruction groups

## 2.1 INTRODUCTION

Single-chip *microcontrollers* are devices designed for use in products that are not usually considered computers, per se, but that require the sophisticated and flexible control that a computer can provide. An example of such a product is an office copier. In contrast to microprocessors such as the 8085, microcontrollers typically integrate RAM, ROM, and I/O, as well as the CPU, onto the same chip. Also, since on-chip program storage (ROM space) is limited on a microcontroller, the instruction set is designed to consist mostly of single-byte instructions.

Many of the applications of microcontrollers fall into one of two categories: *open-loop* or *closed-loop* control systems. Open loop, often called *sequential* control, is used in applications where the process or device being controlled is characterized by a sequence of states, with the progression from state to state being triggered by discrete events. That is, the application is *event driven*. An example is a vending machine that accepts various value coins, recognizes product selection, vends the product, finds the price, and returns the correct change. If the coins were insufficient or the product out of stock, then appropriate messages would be displayed.

Closed-loop control is characterized by the use of *real-time* monitoring of a process to achieve effectively continuous control. The output of the process is monitored using various transducers and A/D converters and the process is continuously modified to achieve the desired result. Examples of closed-loop control can be found in automatic machine tools and robotics.

The use of microcontrollers is not limited to control systems. Other applications include those that require the manipulation of data structures, such as might be found in robot vision or data communications systems.

Microcontrollers are often called *embedded* controllers because they are used as a component of a larger system. The user of the system may not be aware that it contains a processor. On the other hand, users of a stand-alone system, such as a personal computer, are aware that they are running a program even when using the system in a control application. The use of microcontrollers can be very cost-effective, especially in such mass-produced items as microwave ovens, smart modems, and VCRs.

A good example of an advanced 8-bit microcontroller is the Intel MCS-51 family of devices. This family, typified by the 8051, has been designed mainly for sequential control applications. In this chapter we examine the hardware and software features of the 8051 and at points compare them to features in the 8085. However, for a complete and detailed description, the student should obtain and read the Intel literature listed in the references at the end of this chapter.

## 2.2 8051 HARDWARE OVERVIEW

There are three basic members of the MCS-51 family: the 8051, the 8031, and the 8751. The 8051 contains 4K bytes of ROM, 128 bytes of RAM, 32 I/O lines, two 16-bit counter/timers, five interrupt sources (two external), a duplex serial port, and a bit level Boolean processor. The 8031 has no on-board ROM and uses external memory for program storage. The 8751 is the same as the 8051 except that the ROM is replaced by UVEPROM. The 8751 is relatively expensive and is meant to be a program development tool, to be replaced in production by the 8051 containing factory-masked ROM. Three newer devices—the 8052, the 8032, and the 8752—are expanded versions with 8K of ROM, 256 bytes of RAM, and three timers. In addition, there are low-power CMOS versions designated 80C51, 80C31, and 87C51. A block diagram of the 8051/8052 is shown in Fig. 2.1. The 40-pin dual in-line package (DIP) pin connections are shown in Fig. 2.2.

## 2.3 CPU TIMING

### 2.3.1 Clock Frequency

The smallest unit of CPU timing is the oscillator period, also called the *clock*. The 8051 contains an on-chip oscillator that requires a crystal be connected between the pins designated XTAL1 and XTAL2. Alternatively, the 8051 can be driven by an external

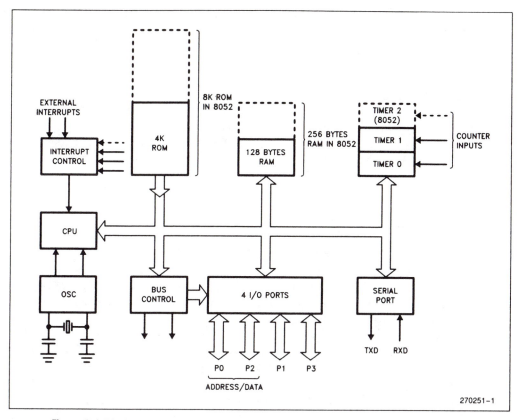

**Figure 2.1** Block diagram of the 8051/8052AH. (*Source:* Reprinted by permission of Intel Corporation, Copyright © Intel Corporation 1987.)

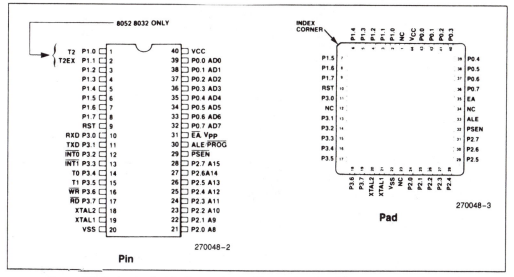

**Figure 2.2** Forty-pin dual in-line package. (*Source:* Reprinted by permission of Intel Corporation, Copyright © Intel Corporation 1987.)

clock signal. For the standard (HMOS) device, the external clock is applied to XTAL2 and XTAL1 is connected to ground ($V_{SS}$). For the CMOS device, external clock is applied to XTAL1 and XTAL2 is left unconnected. The maximum clock frequency is currently 12 MHz.

### 2.3.2 Machine Cycles

A machine cycle consists of six states, designated S1 through S6. Each state is divided into two phases, P1 and P2, each phase lasting one clock period. Thus, a machine cycle is 12 clock periods long, and a 12-MHz clock would give a 1-$\mu$sec machine cycle lasting from S1P1 to S6P2. Many instructions require only one cycle for execution. Refer to Fig. 2.18.

### 2.3.3 Reset

To reset the 8051, a high level must be applied to the RST pin for *at least two* machine cycles. When designing a reset circuit to be activated by a momentary switch or some other short duration signal, the circuit must *stretch out* the signal to last at least 24 clock periods using either an RC time constant or a digital counting scheme. Often a *power-on reset* is implemented by connecting the RST pin to ground through a resistor and connecting a capacitor from RST to the $V_{CC}$ power rail. Typical values are 8.2K and 10 MFD. Such a simple circuit only works under two conditions: $V_{CC}$ comes on with a fast rise time and the clock oscillator starts within a few milliseconds of power on. Because many power supplies have a soft-start feature, the two conditions may not be met and a more complex circuit will be needed for a positive reset.

## 2.4 MEMORY ORGANIZATION

### 2.4.1 Memory Space Separation and Size

In contrast to processors such as the 8085, the 8051 has separate address spaces: one for program storage (usually ROM) and another for data storage (RAM). A given numeric address can refer to two logically and physically different memory locations, depending on the type of instruction using the address. The 8051 supports up to 64K bytes of program storage, the lowest 4K (8K for the 8052) being on chip, the rest external. In addition to on-chip RAM, the 8051 supports up to 64K of external data storage. Note that the 8051 instructions that reference program storage can only read. Even if the physical device used for external program storage is read/write, it is effectively read-only for the 8051. The storage (memory) structure is shown in Fig. 2.3.

### 2.4.2 Program Storage

After reset, the CPU begins fetching instructions starting at address 0000H. The physical location of address 0000H is either on chip or external, depending on the 8051 pin designated $\overline{EA}$ (external address). If $\overline{EA}$ is *low*, address 0000H and all other program

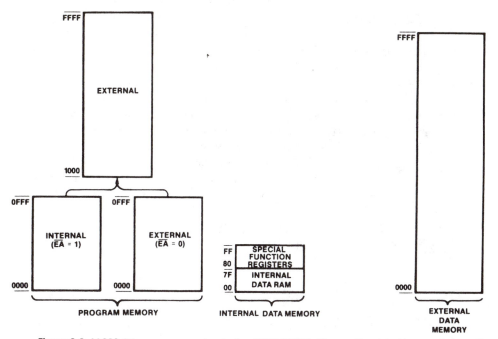

**Figure 2.3** MCS®-51 memory map (excluding 8032/8052). (*Source:* Reprinted by permission of Intel Corporation, Copyright © Intel Corporation 1987.)

storage addresses will reference external memory. If $\overline{EA}$ is *high,* addresses 0000H to 0FFFH (to 1FFFH for the 8052) will reference on-chip ROM; higher address will automatically reference external memory. Note that the 8031 must operate with $\overline{EA}$ connected low because it does not contain any on-chip ROM. As is discussed later, some of the I/O pins are used for address and data when using external memory.

Also, each interrupt is associated with a fixed memory location, the first being address 0003H. An interrupt causes the CPU to jump to that location, which is assumed to be the start of the interrupt service routine. See Fig. 2.4.

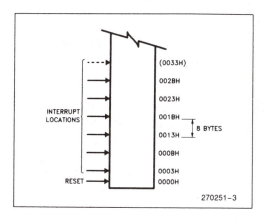

**Figure 2.4** Interrupt vector addresses. (*Source:* Reprinted by permission of Intel Corporation, Copyright © Intel Corporation 1987.)

### 2.4.3 Data Storage

The internal data storage of the 8051 consists of the lower 128 bytes and the SFR (special function registers) space, also 128 bytes. The 8052 contains additional storage called the upper 128 bytes. The lowest 32 bytes of the lower 128 are grouped into four banks of eight registers. Only one bank at a time can be in active use, and it is selected by means of 2 bits in the PSW register that are under program control. The eight registers in the active bank are designated R0 through R7 and can be used by certain software instructions. After reset, the stack pointer (SP) is pointing to the top register of the lowest bank (address 07H). Usually the program loads a new value into SP.

The 16 bytes above the register banks form a block that can be addressed as either bytes or as 128 individual bits. The byte addresses are 20H to 2Fh. The bit addresses are from 00H to 7Fh. Even though the same numeric address (e.g., 20H) can refer to either a byte or a bit, there is no ambiguity. The instruction that uses the address determines if the reference is to be a byte or a bit.

### 2.4.4 On-Chip ROM

The contents of the on-chip ROM in the 8051 are *masked* at the factory. That means that the designer develops a program for the 8051 and sends it to the chip manufacturer. The manufacturer translates the program into a pattern on a photographic mask that is used to define a set of interconnections in the silicon during a final step in chip production. Once the chips are masked, the program cannot be altered.

Three major facts must be understood about masking. First, the system using a mask-programmed part should be stable; that is, it should not require frequent modification of the program. Second, cost savings occur only when enough systems can be sold to pay for the costs associated with developing the mask and making a production run of devices (mass-produced consumer items such as VCRs often qualify). Third, the designer must be absolutely sure the program works and is bug-free before signing off on the mask. Ten thousand bad chips in a box on your desk could ruin your whole day.

The 8031 version of the microcontroller has no on-chip ROM and cannot be masked. Because it uses some of its port pins for the required external program memory, the 8031 does not have the I/O capability of the 8051 unless an external peripheral chip is used. However, for systems with short production runs or frequently updated code the trade-off may be well worth it. Also, the requirement often is to get to market first and lower the production costs later.

The 8751 is the equivalent of the 8051 except that the on-chip ROM is field programmable and erasable with ultraviolet light. The ALE pin on the 8751 also serves as the program pulse input pin during programming, and the $\overline{EA}$ pin is biased to $V_{PP}$, the 21V programming supply. Typically, a special piece of equipment (which may attach to a personal computer) is required to program an 8071. Where the cost of the processor is a small part of the total system cost, it is possible that the 8071 could be the production part.

## 2.5 SFR SPACE: SPECIAL FUNCTION REGISTERS

The registers associated with important functions of the 8051 are assigned memory locations in the on-chip data storage space, allowing them to be addressed by program instructions. Some of the SFR locations are *bit addressable* as well as byte addressable. This is a feature not found in processors such as the 8085. Referring to the SFR map in Fig. 2.5, we see that not all addresses are occupied. In general, unoccupied addresses are not implemented (or at least not documented) and the result of reading or writing to them is indeterminate. Unoccupied addresses in SFR space are reserved for use in future versions of the 8051 and should not be used in programs for the current version.

Table 2.1 shows the contents of the SFR area after reset. Although RAM is not affected by a reset while the 8051 is running, the contents of RAM on power up are indeterminate.

### 2.5.1 Accumulator (ACC) and B Register

When referring to the accumulator as a location in the SFR, the mnemonic ACC is used. Accumulator-specific instructions designate the accumulator as A. The accumulator in the 8051 has the same functions as the accumulator in processors such as the 8085. It is also used in some instructions as an *index register*.

The B register has a specific function in multiply and divide operations. Otherwise, it can be used as a general-purpose scratchpad register.

### 2.5.2 Program Status Word (PSW)

The PSW contains flag bits, as shown in Fig. 2.6. Notice the notation used under the heading "Position" in the figure. A specific bit can be designated by the name of the register followed by a decimal point followed by the number of the bit position within the

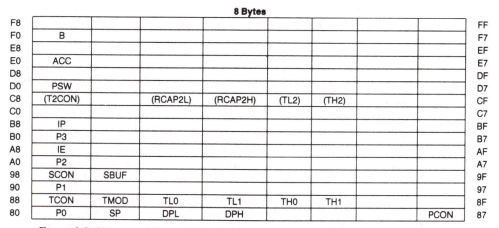

**8 Bytes**

| | | | | | | | | |
|---|---|---|---|---|---|---|---|---|
| **F8** | | | | | | | | **FF** |
| **F0** | B | | | | | | | **F7** |
| **E8** | | | | | | | | **EF** |
| **E0** | ACC | | | | | | | **E7** |
| **D8** | | | | | | | | **DF** |
| **D0** | PSW | | | | | | | **D7** |
| **C8** | (T2CON) | | (RCAP2L) | (RCAP2H) | (TL2) | (TH2) | | **CF** |
| **C0** | | | | | | | | **C7** |
| **B8** | IP | | | | | | | **BF** |
| **B0** | P3 | | | | | | | **B7** |
| **A8** | IE | | | | | | | **AF** |
| **A0** | P2 | | | | | | | **A7** |
| **98** | SCON | SBUF | | | | | | **9F** |
| **90** | P1 | | | | | | | **97** |
| **88** | TCON | TMOD | TL0 | TL1 | TH0 | TH1 | | **8F** |
| **80** | P0 | SP | DPL | DPH | | | | PCON | **87** |

**Figure 2.5** SFR map. (*Source:* Reprinted by permission of Intel Corporation, Copyright © Intel Corporation 1987.)

**Table 2.1** Reset values of the SFRS.

| SFR name | Reset value |
|----------|-------------|
| PC | 0000H |
| ACC | 00H |
| B | 00H |
| PSW | 00H |
| SP | 07H |
| DPTR | 0000H |
| P0–P3 | FFH |
| IP (8051) | XXX00000B |
| IP (8052) | XX000000B |
| IE (8051) | 0XX00000B |
| IE (8052) | 0X000000B |
| TMOD | 00H |
| TCON | 00H |
| TH0 | 00H |
| TL0 | 00H |
| TH1 | 00H |
| TL1 | 00H |
| TH2 (8052) | 00H |
| TL2 (8052) | 00H |
| RCAP2H (8052) | 00H |
| RCAP2L (8052) | 00H |
| SCON | 00H |
| SBUF | Indeterminate |
| PCON (HMOS) | 0XXXXXXXB |
| PCON (CHMOS) | 0XXX0000B |

*Source:* Reprinted by permission of Intel Corporation, Copyright © Intel Corporation 1987.

register. The MSB is position 7 and the LSB is position 0. Thus, the bit designated by the symbol AC can also be designated as PSW.6, as shown. This convention is also followed in the assembly language.

In addition to the usual flags, such as CY and AC, the 8051 has two general-purpose flags not associated with any specific CPU state or function: the bits PSW.5 (symbol F0) and PSW.1 (no symbol). The programmer may use F0 as a flag bit for a user-defined purpose. It may be set and reset by the program as a function of some special condition or be read in from a port pin. The Intel documentation indicates that PSW.1 is reserved for future use and should not be used in programs.

Note that although the PSW does not have a zero flag, this is not a problem because the 8051 has specific instructions to test the accumulator for zero. As mentioned earlier, the PSW also contains two programmable bits (RS0 and RS1) that select which register bank is active.

| (MSB) | | | | | | | (LSB) |
|---|---|---|---|---|---|---|---|
| CY | AC | F0 | RS1 | RS0 | OV | — | P |

| Symbol | Position | Name and Significance | Symbol | Position | Name and Significance |
|---|---|---|---|---|---|
| CY | PSW.7 | Carry flag. | OV | PSW.2 | Overflow flag. |
| AC | PSW.6 | Auxiliary Carry flag. (For BCD operations.) | — | PSW.1 | User definable flag. |
| F0 | PSW.5 | Flag 0 (Available to the user for general purposes.) | P | PSW.0 | Parity flag. Set/cleared by hardware each instruction cycle to indicate an odd/even number of "one" bits in the Accumulator, i.e., even parity. |
| RS1 | PSW.4 | Register bank select control bits 1 & | | | |
| RS0 | PSW.3 | 0. Set/cleared by software to determine working register bank (see Note). | | | |

NOTE:
The contents of (RS1, RS0) enable the working register banks as follows:

(0.0)—Bank 0    (00H–07H)
(0.1)—Bank 1    (08H–0FH)
(1.0)—Bank 2    (10H–17H)
(1.1)—Bank 3    (18H–1FH)

**Figure 2.6** Progam status word register. (*Source:* Reprinted by permission of Intel Corporation, Copyright © Intel Corporation 1987.)

### 2.5.3 Stack Pointer (SP)

Because the stack pointer is 8 bits wide, it allows a maximum stack size of 256 bytes. In contrast to the 8085, the stack in the 8051 grows upward through memory; therefore, the SP is incremented before data are stored as a result of a PUSH or CALL instruction. The stack may reside anywhere in on-chip RAM by loading the appropriate address into the SP. After reset, the SP contains the address 07H, causing the stack to start at location 08H in a register bank. Because the program typically has other uses for the register banks, the stack is usually moved higher in RAM by loading a new address into SP before doing any PUSH or CALL instructions.

### 2.5.4 Data Pointer (DPTR)

The DPTR is a 16-bit quantity held in two 8-bit parts: the high byte in DPH and the low byte in DPL. The main purpose of the DPTR is to hold a 16-bit address for certain instructions. It can be used as a single 16-bit register or as two 8-bit registers.

### 2.5.5 Port Latches (P0, P1, P2, P3)

The 32 I/O pins are organized into four 8-bit ports designated P0–P3. Each port has an associated 8-bit latch, the outputs of which drive the matching I/O pins. The contents of the latches can be read from or written to in the SFR.

### 2.5.6 Serial Data Buffer (SBUF)

The SBUF is actually two separate registers sharing a common address: One is read-only and the other write-only. When data are written to SBUF, they go to a transmit buffer and are held there for serial transmission. When data are read from SBUF, they come from the serial data receive buffer.

### 2.5.7 Timer Registers

Registers TH0 and TL0 are the high and low bytes, respectively, of the 16-bit counting register for timer/counter 0. Likewise, TH1 and TL1 are for timer/counter 1. In the 8052, TH2 and TL2 are for timer/counter 2. Also, the 8052 contains two 8-bit capture registers (RCAP2H and RCAP2L) used to hold copies of the TH2 and TL2 register contents.

### 2.5.8 Control Registers

The SFR contains registers used for the control and status of the interrupt system, the timer/counters, and the serial port. They are IP (interrupt priority), IE (interrupt enable), TMOD (timer mode), TCON (timer control), T2CON (8052 timer 2 control), SCON (serial port control), and PCON (power control, used mainly in 80C51). Each is discussed in a later section.

## 2.6 I/O PORTS

One of the most useful features of the 8051 is the I/O, consisting of four bidirectional ports. Each port has an 8-bit latch in the SFR space, an output driver, and an input buffer. The ports can be used for general I/O, as address and data lines, and for certain special functions.

### 2.6.1 Input, Loading, and Output Drive

Ports 1, 2, and 3 have the equivalent of internal pull-up resistors. When used as inputs, the pins of P1, P2, and P3 will be high (logic 1) when open-circuited and will source current when pulled low by an external device. Port 0 does not have the same pull-up feature and is floating (high impedance) when used as an input. A clarification is needed for the phrase "read a port." Some instructions read the actual level on the I/O pin; others read the level in the latch. A condition can occur where the port is being used to output a high (logic level 1) but the external load attached to the port pin is of such low resistance that the voltage on the pin is at the same level as a low (logic level 0). Reading the latch would show a 1, but reading the pin would show a 0. Instructions that do a read-modify-write operation (e.g., INC) read the latch.

When used as outputs, ports 1, 2, and 3 each can drive the equivalent of four LS TTL inputs. Port 0 can drive eight such equivalent inputs. To speed up 0 to 1 output transitions (i.e., to overcome capacitance effects) on ports 1, 2, and 3, an additional internal pull-up is activated briefly during output.

### 2.6.2 Alternate Port Functions

All the pins of port 3 (and in the 8052, two P1 pins) have an *alternate function,* as listed in Fig. 2.7. To enable an alternate function, a 1 must be written to the corresponding bit in the port latch.

| Port Pin | Alternative Function |
|----------|----------------------|
| P3.0 | RXD (serial input port) |
| P3.1 | TXD (serial output port) |
| P3.2 | $\overline{\text{INT0}}$ (external interrupt 0) |
| P3.3 | $\overline{\text{INT1}}$ (external interrupt 1) |
| P3.4 | T0 (Timer 0 external input) |
| P3.5 | T1 (Timer 1 external input) |
| P3.6 | $\overline{\text{WR}}$ (external data memory write strobe) |
| P3.7 | $\overline{\text{RD}}$ (external data memory read strobe) |

(MSB)                                      (LSB)

| RD | WR | T1 | T0 | INT1 | INT0 | TXD | RXD |
|----|----|----|----|------|------|-----|-----|

| Symbol | Position | Name and Significance |
|--------|----------|------------------------|
| RD | P3.7 | Read data control output. Active low pulse generated by hardware when external data memory is read. |
| WR | P3.6 | Write data control output. Active low pulse generated by hardware when external data memory is written. |
| T1 | P3.5 | Timer/counter 1 external input or test pin. |
| T0 | P3.4 | Timer/counter 0 external input or test pin. |
| INT1 | P3.3 | Interrupt 1 input pin. Low-level or falling-edge triggered. |
| INT0 | P3.2 | Interrupt 0 input pin. Low-level or falling-edge triggered. |
| TXD | P3.1 | Transmit Data pin for serial port in UART mode. Clock output in shift register mode. |
| RXD | P3.0 | Receive Data pin for serial port in UART mode. Data I/O pin in shift register mode. |

**Figure 2.7** Alternate functions of port 3 pins. (*Source:* Reprinted by permission of Intel Corporation, Copyright © Intel Corporation 1987.)

### 2.6.3 Accessing External Memory

Because the 8051 has separate program memory and data memory, it uses different hardware signals to access the corresponding external storage devices. The PSEN (program store enable) signal is used as the read strobe for program memory, and RD and WR are used as the read and write strobes to access data memory. Note that RD and WR are alternate functions of the P3.6 and P3.7 pins, as described above. Also, ports 0 and 2 are used to access external memory.

Accesses to external program store always use a 16-bit address. Accesses to external data store may use either an 8-bit or a 16-bit address, depending on the instruction being executed. In the case of a 16-bit address, the high-order 8 bits of the address are output on port 2, where they are held constant during the entire memory access cycle. The prior contents of the port 2 latches in the SFR are not lost but are restored after the memory access cycle. If an 8-bit address is being used, the contents of port 2 are unchanged, which allows some of the port 2 pins to be used to select 256-byte pages for the lower 8 bits of the address.

The low-order 8 bits of the address are multiplexed with the data byte on port 0. When used in this mode, the port 0 pins are connected to an internal active pull-up; they do not float. The prior contents of the port 0 latches are lost. The ALE (address latch enable) signal must be used to capture the low-order address bits in an external latch, much the same as is done with the 8085.

## 2.7 TIMER/COUNTERS

The 8051 has two 16-bit registers that can be used as either timers or counters. They are designated timer 0 and timer 1. The 8052 has an additional 16-bit register designated timer 2. These registers are in the SFR as pairs of 8-bit registers.

When used as a timer, the register is incremented once per machine cycle, which is equal to once per 12 clock periods. When used as a counter, the register is incremented on a 1-0 transition (a negative edge) applied to the appropriate input pin: T0 or T1 (or T2 in the 8052). Remember that T0, T1, and T2 are alternate functions of port pins (refer to Fig. 2.7). It takes two complete machine cycles for the 8051 to see the 1-0 transition; the input must be held high for at least one cycle and then low for at least one cycle.

### 2.7.1 Timer 0 and Timer 1

The way that timer 0 and timer 1 will operate is determined by the 8 bits written to the TMOD register, as detailed in Fig. 2.8a. The bits M0 and M1 (a pair for each timer) are used to select one of four operating modes: mode 0, mode 1, mode 2, or mode 3. Both counters work the same in modes 0, 1, and 2 but differently in mode 3.

### 2.7.2 Mode 0 and Mode 1

In mode 0 the timer is configured as a 13-bit counter that can be thought of as an 8-bit counter preceded by a 5-bit divide-by-32 prescaler. The 8-bit count is in the TH register (TH0 or TH1, depending on which counter is in use), and the 5-bit prescaler is the lower 5 bits of the TL register. The upper 3 bits of the TL register are random and should be ignored. As the 13-bit count in TL and TH goes from all 1s to all 0s, the timer interrupt flag (TF0 or TF1) is set. The timer interrupt is the connection between the counter hardware and the program software.

As in Fig. 2.9, the source of input to the counter is selected by the C/T bit in TMOD. The counting process can be turned on and off (enabled or disabled) independent of the input. In order for counting to proceed, the TR bit (TR0 or TR1) in the TCON register (Fig. 2.8b) must be a 1 at the same time that one of the following is true: Either the appropriate GATE bit in the TMOD register is a 0, or the appropriate INT pin (INT0 or INT1) is held low. The use of GATE or TR allows counting to be controlled by *software;* the use of INT allows counting to be controlled by external hardware. Remember that INT0 and INT1, as well as T0 and T1 inputs, are alternate functions of port 3 pins. Mode 1 is the same as mode 0, except that the timer register is 16 bits long, with all 8 bits of TL being used.

| (MSB) | | | | | | | (LSB) |
|---|---|---|---|---|---|---|---|
| GATE | C/T̄ | M1 | M0 | GATE | C/T̄ | M1 | M0 |

**Timer 1**                    **Timer 0**

| | | M1 | M0 | Operating Mode |
|---|---|---|---|---|
| GATE | Gating control when set. Timer/Counter "x" is enabled only while "INTx" pin is high and "TRx" control pin is set. When cleared Timer "x" is enabled whenever "TRx" control bit is set. | 0 | 0 | MCS-48 Timer "TLx" serves as 5-bit prescaler. |
| | | 0 | 1 | 16-bit Timer/Counter "THx" and "TLx" are cascaded; there is no prescaler. |
| C/T̄ | Timer or Counter Selector cleared for Timer operation (input from internal system clock). Set for Counter operation (input from "Tx" input pin). | 1 | 0 | 8-bit auto-reload Timer/Counter "THx" holds a value which is to be reloaded into "TLx" each time it overflows. |
| | | 1 | 1 | (Timer 0) TL0 is an 8-bit Timer/Counter controlled by the standard Timer 0 control bits. TH0 is an 8-bit timer only controlled by Timer 1 control bits. |
| | | 1 | 1 | (Timer 1) Timer/Counter 1 stopped. |

**Figure 2.8a** Timer/counter mode control register. (*Source:* Reprinted by permission of Intel Corporation, Copyright © Intel Corporation 1987.)

| (MSB) | | | | | | | (LSB) |
|---|---|---|---|---|---|---|---|
| TF1 | TR1 | TF0 | TR0 | IE1 | IT1 | IE0 | IT0 |

| Symbol | Position | Name and Significance | Symbol | Position | Name and Significance |
|---|---|---|---|---|---|
| TF1 | TCON.7 | Timer 1 overflow Flag. Set by hardware on Timer/Counter overflow. Cleared by hardware when processor vectors to interrupt routine. | IE1 | TCON.3 | Interrupt 1 Edge flag. Set by hardware when external interrupt edge detected. Cleared when interrupt processed. |
| TR1 | TCON.6 | Timer 1 Run control bit. Set/cleared by software to turn Timer/Counter on/off. | IT1 | TCON.2 | Interrupt 1 Type control bit. Set/cleared by software to specify falling edge/low level triggered external interrupts. |
| TF0 | TCON.5 | Timer 0 overflow Flag. Set by hardware on Timer/Counter overflow. Cleared by hardware when processor vectors to interrupt routine. | IE0 | TCON.1 | Interrupt 0 Edge flag. Set by hardware when external interrupt edge detected. Cleared when interrupt processed. |
| TR0 | TCON.4 | Timer 0 Run control bit. Set/cleared by software to turn Timer/Counter on/off. | IT0 | TCON.0 | Interrupt 0 Type control bit. Set/cleared by software to specify falling edge/low level triggered external interrupts. |

**Figure 2.8b** Timer/counter control register. (*Source:* Reprinted by permission of Intel Corporation, Copyright © Intel Corporation 1987.)

### 2.7.3 Mode 2

Mode 2 operation is much the same as mode 0, except that the TL register is used as an 8-bit counter and the TH register is used to hold a preset number. When the contents of TL go from all 1s to all 0s, the interrupt (TF) is set and the contents of TH are transferred to TL. The contents of TH remain unchanged. Because the contents of TH are under software control, the counter can be made to divide the count source by any number from 1 to 255 by means of the automatic reload of TH into TL.

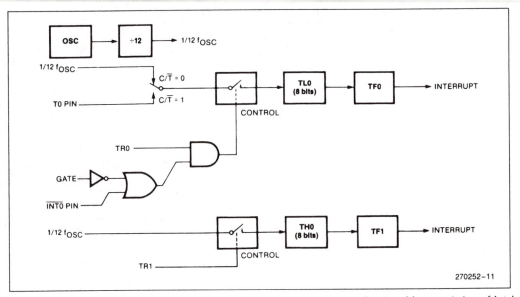

**Figure 2.9** Timer/counter 0 mode 3: two 8-bit counters. (*Source:* Reprinted by permission of Intel Corporation, Copyright © Intel Corporation 1987.)

### 2.7.4 Mode 3

In mode 3, timer 1 is disabled but holds its count; it is essentially frozen. Timer 0 in mode 3 is split into two separate counters (see Fig. 2.9). The first counter is the same as mode 0, except TL0 is used as an 8-bit counter and there is no prescaler. The second counter uses TH0 as an 8-bit counter. The count source is the oscillator divided by 12 and the enable control is the TR1 bit in the TCON register. Note that the first counter sets the TF0 interrupt flag and the second counter sets TF1.

### 2.7.5 Timer 2

Timer 2 is a 16-bit timer/counter in the 8052 group of devices; it does not appear in the 8051 group. Like the timers already described, the input source for timer 2 can be the clock (timer operation) or an external input (counter operation). Timer 2 has three operating modes: capture, auto-load, and baud rate generator. The input source and operating mode are selected by bits in the T2CON control register. A detailed description of timer 2 can be found in the Intel literature listed at the end of this chapter.

## 2.8 SERIAL PORT INTERFACE

The 8051 has a full duplex serial port that allows data to be transmitted and received simultaneously in hardware while the software is doing other things. A serial port interrupt is generated by the hardware to get the attention of the program in order to read

Chap. 2: The 8051 Single-Chip Microcontroller

or write serial port data. The receiver hardware is *double buffered,* meaning that a received frame of data can be held for reading while a second frame is being received. Double buffering allows the receiver interrupt service routine to be less time critical, but the stored frame must be read before reception of the second frame is complete or the stored frame will be overwritten and lost. To obtain the same level of performance with a processor such as the 8085, the use of a separate USART (universal synchronous/ asynchronous receiver/transmitter) chip is required.

As mentioned, both the transmit and receive buffers are accessed at the same location in the SFR space: the SBUF register. Writing to SBUF loads the transmit buffer, and reading from SBUF obtains the contents of the receive buffer. Any instruction that writes to SBUF initiates serial transmission. The serial port has four modes of operation: mode 0, mode 1, mode 2, and mode 3. (Note that these modes are not to be confused with the timer/counter modes; a designation such as ''mode 1'' must be understood in the context of the hardware feature being discussed.)

### 2.8.1 Serial Port Control Register (SCON)

SCON is the serial port control and status register (see Fig. 2.10). Bits SM0 and SM1 are used to select the operating mode. The SM2 bit is used in a multiprocessor system where one 8051 acts as a master unit, sending commands to one or more slave units. Such systems are beyond the scope of this book, and the reader is referred to the Intel documentation for more information.

|  | (MSB) | | | | | | | (LSB) |
|---|---|---|---|---|---|---|---|---|
|  | SM0 | SM1 | SM2 | REN | TB8 | RB8 | TI | RI |

Where SM0, SM1 specify the serial port mode, as follows:

| SM0 | SM1 | Mode | Description | Baud Rate |
|---|---|---|---|---|
| 0 | 0 | 0 | shift register | $f_{osc.}/12$ |
| 0 | 1 | 1 | 8-bit UART | variable |
| 1 | 0 | 2 | 9-bit UART | $f_{osc.}/64$ or $f_{osc.}/32$ |
| 1 | 1 | 3 | 9-bit UART variable | |

- SM2    enables the multiprocessor communication feature in Modes 2 and 3. In Mode 2 or 3, if SM2 is set to 1 then RI will not be activated if the received 9th data bit (RB8) is 0. In Mode 1, if SM2 = 1 then RI will not be activated if a valid stop bit was not received. In Mode 0, SM2 should be 0.

- REN    enables serial reception. Set by software to enable reception. Clear by software to disable reception.

- TB8    is the 9th data bit that will be transmitted in Modes 2 and 3. Set or clear by software as desired.

- RB8    in Modes 2 and 3, is the 9th data bit that was received. In Mode 1, if SM2 = 0, RB8 is the stop bit that was received. In Mode 0, RB8 is not used.

- TI    is transmit interrupt flag. Set by hardware at the end of the 8th bit time in Mode 0, or at the beginning of the stop bit in the other modes, in any serial transmission. Must be cleared by software.

- RI    is receive interrupt flag. Set by hardware at the end of the 8th bit time in Mode 0, or halfway through the stop bit time in the other modes, in any serial reception (except see SM2). Must be cleared by software.

**Figure 2.10** Serial port control and status register. (*Source:* Reprinted by permission of Intel Corporation, Copyright © Intel Corporation 1987.)

## 2.8.2 Serial Port Modes

Mode 0 is *half-duplex synchronous* operation. Data are sent and received (but not simultaneously) through the RXD pin in 8-bit frames, LSB first. The bit rate is fixed at one-twelfth the oscillator frequency. The shift clock, which is the same frequency as the bit rate, is sent out the TXD pin during both transmission and reception and is used to synchronize the receiver to the sender. A shift clock edge will occur during the valid state of each data bit. Note that TXD and RXD are alternate functions of port 3 pins. In mode 0, reception is initiated when bit REN is set to 1 and bit RI is cleared to 0 in the SCON register.

Mode 1 is *full duplex asynchronous* operation. Data are sent out TXD and received through RXD. A complete frame consists of a start bit (always a 0), followed by 8 data bits (LSB first), followed by a stop bit (always a 1). The start and stop bits are added by the hardware; the software writes the 8-bit data byte to, or reads it from, SBUF. The baud rate is variable and can be obtained by using timer 1 as a baud rate generator, as described in Sec. 2.8.3.

Mode 2 is similar to mode 1, with two exceptions. First, the frame is 11 bits long; a ninth data bit is inserted before the stop bit. When transmitting, the ninth bit is obtained from TB8 in SCON. It is assumed that TB8 was written into before initiating transmission. When receiving, the ninth bit can be read from RB8 in SCON. A common use for the ninth bit is as a parity bit for 8-bit data. The second difference is that the baud rate is either 1/32nd or 1/64th of the oscillator frequency, as selected by the SMOD bit (bit 7) in the PCON register. If SMOD is set to 1, the 1/32 number is used. If SMOD is cleared to 0, 1/64 is used.

Mode 3 is the same as mode 2, except that the baud rate is variable and can be obtained in the same way as in mode 1.

Reception for modes 1, 2, and 3 is enabled when the REN bit in SCON is set to 1. Actual reception is initiated when an incoming start bit causes a high to low transition on the RXD pin. As mentioned earlier, transmission is initiated by writing to SBUF in any serial mode.

## 2.8.3 Timer 1 as Baud Rate Generator

Although it is possible to use the timer/counter as a baud rate generator in any of its modes, its most common use is as a timer (i.e., clock-sourced) in auto-reload mode (timer mode 2). The baud rate is then given by the expression

$$\text{Baud rate} = \frac{\text{Oscillator frequency}}{N[256 - (\text{TH1})]}$$

where $N$ depends on the SMOD bit in the PCON register (bit PCON.7). If SMOD = 0, then $N = 384$. If SMOD = 1, then $N = 192$. The term (TH1) represents the contents of register TH1.

When selecting a crystal to set the oscillator frequency in a system that will use the serial port, consider the following. Because the value stored in TH1 must, by definition, be an integer number, the oscillator frequency must be an integer multiple of both the baud rate and $N$ in order to obtain an exact baud rate using the above expression.

---

**EXAMPLE 2.1 CRYSTAL SELECTION**

**PROBLEM**   Pick a crystal frequency close to 12 MHz that will allow baud rates of 300, 600, and 1200 bits/sec.

**SOLUTION**   Assume we set SMOD so that $N = 384$. Find a value for (TH1) in the range 0–255 for one baud rate and check it for the others:

$$1200 \times 384 \times [256 - (TH1)] = 12 \times 10^6 \quad \text{(approximately)}$$

$$(TH1) = 256 - (12 \times 10^6)/(384 \times 1200)$$

$$(TH1) = 256 - 26.04 = 230 \quad \text{(must be an integer)}$$

Next, use $(TH1) = 230$ to find the exact crystal frequency:

$$F = 1200 \times 384 \times 26 = 11.981 \text{ MHz} \quad \text{(to 3 decimal points)}$$

Next, verify other baud rates (note the baud rate ratios).

$$\text{For 600 baud: } 256 - (TH1) = 2 \times 26 = 52, \text{ so } (TH1) = 204$$

$$(11.981 \times 10^6)/(384 \times 52) = 600 \quad \textit{verified}$$

$$\text{For 300 baud: } 256 - (TH1) = 2 \times 52 = 104, \text{ so } (TH1) = 152$$

$$(11.981 \times 10^6)/(384 \times 104) = 300 \quad \textit{verified}$$

---

### 2.8.4 The PCON Register

Of the 8 bits in the PCON power control register, only bit 7, SMOD, is implemented in the standard 8051. As discussed, SMOD is used in setting the baud rate of the serial port. Bits 0, 1, 2, and 3 are implemented in the CMOS version. Bits 0 and 1 are used in power-saving modes and bits 2 and 3 are general-purpose flags. Details of the use of PCON in the CMOS devices can be found in the Intel literature.

## 2.9 INTERRUPTS

The 8051 has five sources of interrupts: two from external pins (INT0 and INT1), two from the timer/counters (TF0 and TF1), and one from the serial port (TI or SI). A useful feature of the 8051 is that the interrupt sources are associated with bit locations in registers. Those bits can be set or cleared by software, with the same results as when the bits are set or cleared by hardware.

As shown in Fig. 2.11, all interrupts or each individual interrupt can be enabled or disabled by setting or clearing the appropriate bit in the IE register. If enabled, an interrupt will cause a call to one of the predefined locations in RAM, as discussed in section 2.4.2 The return address is automatically pushed onto the stack before jumping and is popped back off when an RETI (return-from-interrupt) instruction is executed. If an interrupt

```
         (MSB)                              (LSB)
        | EA | — | ET2 | ES | ET1 | EX1 | ET0 | EX0 |
```

| Symbol | Position | Function |
|--------|----------|----------|
| EA | IE.7 | disables all interrupts. If EA = 0, no interrupt will be acknowledged. If EA = 1, each interrupt source is individually enabled or disabled by setting or clearing its enable bit. |
| — | IE.6 | reserved. |
| ET2 | IE.5 | enables or disables the Timer 2 Overflow or capture interrupt. If ET2 = 0, the Timer 2 interrupt is disabled. |
| ES | IE.4 | enables or disables the Serial Port interrupt. If ES = 0, the Serial Port interrupt is disabled. |
| ET1 | IE.3 | enables or disables the Timer 1 Overflow interrupt. If ET1 = 0, the Timer 1 interrupt is disabled. |
| EX1 | IE.2 | enables or disables External Interrupt 1. If EX1 = 0, External Interrupt 1 is disabled. |
| ET0 | IE.1 | enables or disables the Timer 0 Overflow interrupt. If ET0 = 0, the Timer 0 interrupt is disabled. |
| EX0 | IE.0 | enables or disables External Interrupt 0. If EX0 = 0, External Interrupt 0 is disabled. |

User software should never write 1s to unimplemented bits, since they may be used in future MCS-51 products.

**Figure 2.11** Interrupt enable register. (*Source:* Reprinted by permission of Intel Corporation, Copyright © Intel Corporation 1987.)

occurs while it is disabled, or while a higher priority one is running, it becomes *pending*. As soon as a pending interrupt is enabled, it will cause a call, unless it was cancelled by software while it was still pending or a higher priority interrupt was simultaneously made active.

### 2.9.1 Servicing External Devices

The timing of events in external devices usually has no relationship to the timing of the CPU; in other words, real-world events are usually asynchronous to the processor. In order to monitor and control external devices, a microcomputer must have a method of responding to I/O requests and other external events in a timely manner.

### 2.9.2 Polling and Buffering

One way a processor deals with I/O devices is to ask them periodically if they need service; that is, to poll them. Often, a polled device will exchange blocks of information with the processor. The device will hold such blocks in its own memory, the buffer. The main drawback to polling is the amount of time the CPU spends checking the I/O device.

If the I/O device buffer either fills up or empties out while it is waiting for service, data may be lost or the device may stop working. Consider a printer. The processor initially will load the print buffer with text. But if the buffer runs out before the printer is polled again, printing will stop. Another example is digital communications equipment. If data keep coming in, the receiver buffer can overflow while waiting for the next poll.

Also, real-time applications, which require service from the processor as soon as the need occurs, cannot wait for a poll. Clearly, something better is called for. That something is an interrupt.

### 2.9.3 Basic Interrupt Action

I/O devices often require immediate service while the processor is in the middle of doing something else. The interrupt is a software-controlled hardware feature that, in an orderly manner, forces the processor to suspend what it is currently doing in order to service the I/O device. When it is finished with I/O, the processor will resume where it left off, much the same as a subroutine. Interrupts that can be blocked by software are called *maskable;* those that cannot be blocked are called *nonmaskable*. Interrupts that occur while masked are said to be *pending*.

An I/O device will request service by activating an interrupt pin on the CPU. If the CPU has enabled its interrupts in software, it will initiate its response, often with an acknowledgment signal to the I/O device. This request-acknowledge sequence is an example of handshaking.

The rest of the response is similar to a subroutine call. The CPU will push the return address onto the stack and branch to a predefined part of memory, where it expects to find the *interrupt service routine* which will handle the interrupting device. The last instruction in the routine will be a RETURN, which will pop the return address off the stack and into the PC register. The CPU will then resume program execution from the point where it was interrupted.

### 2.9.4 Multiple Interrupt Sources and Vectoring

It is possible to have more than one interrupt source. When an interrupt occurs, the processor must first determine which device was the cause, as different devices need different services. One way is for the interrupt service routine to poll all the devices to find out which one called, perhaps by having the service routine read status lines from all the devices. But polling can be time-consuming. A faster way is for each interrupting device to point (like a vector arrow) to the place in memory where its service routine is stored. The CPU can then go there directly. Such a system is called a *vectored interrupt*.

In a vectored interrupt system, a problem can arise if two devices request an interrupt at exactly the same time. Which gets serviced first? What is needed is a means of establishing priority. As we shall see, the 8051 has a priority scheme.

### 2.9.5 Priority Levels

The 8051 has a *two-tier* priority structure. The top tier has two priority levels: high and low. Each interrupt source can be assigned to either high-level or low-level status by setting the appropriate bits in the IP register, as shown in Fig. 2.12. When two interrupts of different levels are received simultaneously, the high-level interrupt is serviced first. The second tier of priority is used to resolve simultaneous interrupts within the same level. The *priority-within-level* ordering from highest to lowest is fixed as follows: IE0, TF0, IE1, TF1, RI, or TI.

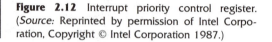

| (MSB) | | | | | | (LSB) |
|---|---|---|---|---|---|---|
| — | — | PT2 | PS | PT1 | PX1 | PT0 | PX0 |

| Symbol | Position | Function |
|---|---|---|
| — | IP.7 | reserved |
| — | IP.6 | reserved |
| PT2 | IP.5 | defines the Timer 2 interrupt priority level. PT2 = 1 programs it to the higher priority level. |
| PS | IP.4 | defines the Serial Port interrupt priority level. PS = 1 programs it to the higher priority level. |
| PT1 | IP.3 | defines the Timer 1 interrupt priority level. PT1 = 1 programs it to the higher priority level. |
| PT0 | IP.1 | defines the Timer 0 interrupt priority level. PT0 = 1 programs it to the higher priority level. |
| PX0 | IP.0 | defines the External Interrupt 0 priority level. PX0 = 1 programs it to the higher priority level. |

User software should never write 1s to unimplemented bits, since they may be used in future MCS-51 products.

**Figure 2.12** Interrupt priority control register. (*Source:* Reprinted by permission of Intel Corporation, Copyright © Intel Corporation 1987.)

Note that bits 7, 6, and 5 in the IP register are unimplemented in the 8051 and should not be used by the software. Bit 5 does have a use in the 8052.

### 2.9.6 Interrupt Timing and Handling

As shown in Fig. 2.13, a "snap-shot" (sample) of the interrupt flags is taken by the 8051 hardware at the end of a typical machine cycle (C1). During the following machine cycle (C2), the sample from the previous cycle is examined. If one of the flags sampled during C1 is found to be set, then a call to the appropriate interrupt vector will be generated during cycles C3 and C4. Execution of the interrupt service routine will start with cycle C5 and continue for as long as is required by the routine.

The hardware will not generate the interrupt call if one of the following is true:

1. An interrupt of equal or higher priority is *already in progress*. A lower priority interrupt can itself be interrupted by a higher priority interrupt, assuming the higher one is enabled.

2. The current machine cycle is *not the final cycle* of the instruction being executed. The instruction in progress must be completed before jumping to the interrupt service instructions.

3. The instruction in progress is RETI or any instruction that writes to the IE or IP register. At least one *additional instruction* must be executed following those before jumping to the interrupt vector address. The reason for this condition is to prevent an interrupt from occurring in the middle of a routine that is in the process of reconfiguring the interrupts. Interrupt-driven applications are sufficiently complex without the added chaos of not being able to gain access to the interrupt control registers due to repeated interrupts.

Chap. 2: The 8051 Single-Chip Microcontroller

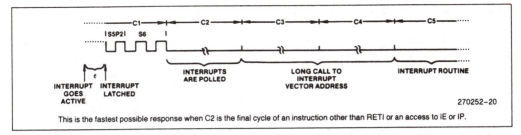

This is the fastest possible response when C2 is the final cycle of an instruction other than RETI or an access to IE or IP.

270252-20

**Figure 2.13** Interrupt response timing diagram. (*Source:* Reprinted by permission of Intel Corporation, Copyright © Intel Corporation 1987.)

The time between the activation of an interrupt and the start of execution of the service routine is the *response time*. In the 8051, the shortest response time is three machine cycles and the longest (worst case) is nine machine cycles. In time-critical applications, the worst case condition must be assumed to happen.

### 2.9.7 Activation Levels and Flag Clearing

In general, interrupts can be either *level-activated* or *transition-activated* (level-triggered or edge-triggered). Because a transition-activated event is, by definition, a transient, when the flag bit associated with the interrupt is cleared, the interrupt event itself is cleared. On the other hand, clearing the flag bit of a level-activated interrupt will have no effect if the external level causing the interrupt stays active.

In the 8051, the external interrupts INT0 and INT1 can be individually configured to be either transition activated or level activated, depending on the value of the bits IT0 and IT1 in the TCON register. The flag bit will be set when the interrupt occurs and cleared automatically when the call is made to the interrupt vector.

For the timers, the TF0 and TF1 flags are set when the count in the corresponding counter register rolls over from all 1s to all 0s; the counter interrupts are in effect transition activated. TF0 and TF1 are also cleared automatically during interrupt service.

The serial port interrupt is generated when either the RI bit or the TI bit is set to 1. However, RI and TI are not automatically cleared. The interrupt service routine will have to determine which bit caused the interrupt and then clear it as part of the routine.

### 2.9.8 Predefined Vector Addresses

As shown in Figure 2.4, the 8051 has predefined vector addresses for interrupts. An 8051 assembler will typically recognize the following predefined code addresses:

| Symbol | Address | Interrupt Source |
|--------|---------|------------------|
| RESET | 00H | Power Up or Reset |
| EXTI0 | 03H | External Interrupt 0 |
| TIMER0 | 0BH | Timer 0 Interrupt |
| EXTI1 | 13H | External Interrupt 1 |
| TIMER1 | 1BH | Timer 1 Interrupt |
| SINT | 23H | Serial Port Interrupt |

Sec. 2.9: Interrupts

**39**

## 2.10 INSTRUCTIONS AND ADDRESSING

The programming model for the 8085 is essentially the accumulator, the flags, a few registers, and a relatively simple vectored interrupt system. In contrast, the programming model for the 8051 has taken the first nine sections of this chapter to describe. Likewise, the assembly language of the 8051 contains features, such as Boolean operations, not seen in similar processors but that give the 8051 its flexibility and power as a control device. The language, however, is not complex.

The 8051 has five addressing modes and five groups of instructions. Each is described in the following sections. The complete instruction set is described in Appendix C. A summary of the instruction set is at the end of this chapter.

### 2.10.1 Direct Addressing

Instructions using direct addressing are 2 bytes long: an 8-bit op-code followed by an 8-bit address. The address is a location in internal RAM or in the SFR area. Depending on the type of instruction, the address refers to either a byte location or a specific bit in a bit-addressable byte.

**EXAMPLE 2.2 SOME DIRECT INSTRUCTIONS**

**MOV A,07;**  Move contents of RAM *byte* location 07 to the accumulator.
**CLR 07;**  Clear bit address 07, the MSB of byte address 20H.

Figure 2.14 shows the bit-addressable bytes in RAM together with the bit-addressable bytes in the SFR area. Assemblers for the 8051 usually have predefined mnemonic symbols corresponding to important bit and byte addresses. Table 2.2 shows the bit address names, and Table 2.3 shows the byte address names.

### 2.10.2 Indirect Addressing

An instruction using direct addressing specifies the fixed address of the operand, an instruction using indirect addressing specifies a register that contains the address of the operand. Because the contents of the register can be changed by the program, indirect addressing is a powerful technique. Also, most of the instructions that use indirect addressing are only 1 byte long, making them efficient in memory space and execution time.

Indirect addressing can access external RAM as well as internal, and addresses can be 8 bits or 16 bits. The 8-bit addresses use the registers designated R0–R7; the 16-bit addresses use the DPTR register. Remember that DPTR is actually the two 8-bit registers DPH and DPL treated as one 16-bit register. Also remember that different banks of registers can be selected for R0–R7, depending on bits in the PSW.

**EXAMPLE 2.3 SOME INDIRECT INSTRUCTIONS**

**MOV A,@R3;**  Move to the accumulator the internal RAM byte pointed to by register R3.
**MOVX A,@DPTR;**  Move to the accumulator the external RAM byte pointed to by the DPTR register.

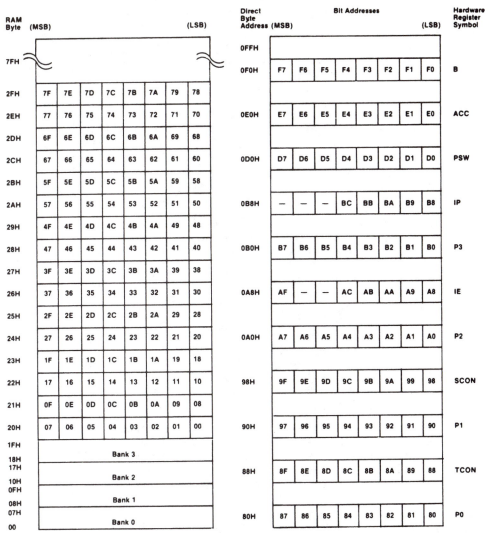

**Figure 2.14** (a) RAM bit addresses. (b) Special function register bit addresses. (*Source:* Reprinted by permission of Intel Corporation, Copyright © Intel Corporation 1987.)

### 2.10.3 Register Instructions

Register instructions use the contents of one of the registers, typically R0–R7, as the operand. All register instructions have the efficiency of being 1 byte long. A few register instructions are register specific; they do not allow the programmer to specify the operand register.

> **EXAMPLE 2.4 SOME REGISTER INSTRUCTIONS**
>
> **MOV A,R3;**   Move contents of register R3 to the accumulator.
> **MUL AB;**   Multiply the contents of accumulator by the contents of the B register.

Table 2.2 Predefined bit addresses for 8051.

| Symbol | Bit position | Bit address | Meaning |
|--------|--------------|-------------|---------|
| CY | PSW.7 | D7H | Carry Flag |
| AC | PSW.6 | D6H | Auxiliary Carry Flag |
| F0 | PSW.5 | D5H | Flag 0 |
| RS1 | PSW.4 | D4H | Register Bank Select Bit 1 |
| RS0 | PSW.3 | D3H | Register Bank Select Bit 0 |
| OV | PSW.2 | D2H | Overflow Flag |
| P | PSW.0 | D0H | Parity Flag |
| TF1 | TCON.7 | 8FH | Timer 1 Overflow Flag |
| TR1 | TCON.6 | 8EH | Timer 1 Run Control Bit |
| TF0 | TCON.5 | 8DH | Timer 0 Overflow Flag |
| TR0 | TCON.4 | 8CH | Timer 0 Run Control Bit |
| IE1 | TCON.3 | 8BH | Interrupt 1 Edge Flag |
| IT1 | TCON.2 | 8AH | Interrupt 1 Type Control Bit |
| IE0 | TCON.1 | 89H | Interrupt 0 Edge Flag |
| IT0 | TCON.0 | 88H | Interrupt 0 Type Control Bit |
| SM0 | SCON.7 | 9FH | Serial Mode Control Bit 0 |
| SM1 | SCON.6 | 9EH | Serial Mode Control Bit 1 |
| SM2 | SCON.5 | 9DH | Serial Mode Control Bit 2 |
| REN | SCON.4 | 9CH | Receiver Enable |
| TB8 | SCON.3 | 9BH | Transmit Bit 8 |
| RB8 | SCON.2 | 9AH | Receive Bit 8 |
| TI | SCON.1 | 99H | Transmit Interrupt Flag |
| RI | SCON.0 | 98H | Receive Interrupt Flag |
| EA | IE.7 | AFH | Enable All Interrupts |
| ES | IE.4 | ACH | Enable Serial Port Interrupt |
| ET1 | IE.3 | ABH | Enable Timer 1 Interrupt |
| EX1 | IE.2 | AAH | Enable External Interrupt 1 |
| ET0 | IE.1 | A9H | Enable Timer 0 Interrupt |
| EX0 | IE.0 | A8H | Enable External Interrupt 0 |
| RD | P3.7 | B7H | Read Data for External Memory |
| WR | P3.6 | B6H | Write Data for External Memory |
| T1 | P3.5 | B5H | Timer/Counter 1 External Flag |
| T0 | P3.4 | B4H | Timer/Counter 0 External Flag |
| INT1 | P3.3 | B3H | Interrupt 1 Input Pin |
| INT0 | P3.2 | B2H | Interrupt 0 Input Pin |
| TXD | P3.1 | B1H | Serial Port Transmit Pin |
| RXD | P3.0 | B0H | Serial Port Receive Pin |
| PS | IP.4 | BCH | Priority of Serial Port Interrupt |
| PT1 | IP.3 | BBH | Priority of Timer 1 Interrupt |
| PX1 | IP.2 | BAH | Priority of External Interrupt 1 |
| PT0 | IP.1 | B9H | Priority of Timer 0 |
| PX0 | IP.0 | B8H | Priority of External Interrupt 0 |

*Source:* Reprinted by permission of Intel Corporation, Copyright © Intel Corporation 1987.

**Table 2.3** Predefined data addresses for 8051.

| Symbol | Hexadecimal address | Meaning |
|--------|--------------------|---------|
| ACC  | E0 | Accumulator |
| B    | F0 | Multiplication Register |
| DPH  | 83 | Data Pointer (high byte) |
| DPL  | 82 | Data Pointer (low byte) |
| IE   | A8 | Interrupt Enable |
| IP   | B8 | Interrupt Priority |
| P0   | 80 | Port 0 |
| P1   | 90 | Port 1 |
| P2   | A0 | Port 2 |
| P3   | B0 | Port 3 |
| PSW  | D0 | Program Status Word |
| SBUF | 99 | Serial Port Buffer |
| SCON | 98 | Serial Port Controller |
| SP   | 81 | Stack Pointer |
| TCON | 88 | Timer Control |
| TH0  | 8C | Timer 0 (high byte) |
| TH1  | 8D | Timer 1 (high byte) |
| TL0  | 8A | Timer 0 (low byte) |
| TL1  | 8B | Timer 1 (low byte) |
| TMOD | 89 | Timer Mode |

*Source:* Reprinted by permission of Intel Corporation, Copyright © Intel Corporation 1987.

### 2.10.4 Immediate Operand Instructions

Instructions using an immediate operand have a numeric constant (or its symbolic name) following the op-code. The constant can be 8 or 16 bits long, depending on the instruction.

> **EXAMPLE 2.5 SOME IMMEDIATE INSTRUCTIONS**
>
> **MOV A,#NUM8;**        Move the 8-bit number NUM8 to the accumulator.
> **MOV DPTR,#NUM16;**  Move the 16-bit number NUM16 to the data pointer.

### 2.10.5 Indexed Addressing

The 8051 has two uses for indexed addressing: reading data tables from program memory space and implementing jump tables. In either case, a 16-bit register holds a base address and the accumulator holds an 8-bit displacement or index. The address of the data byte (or the jump address, as the case may be) is the sum of the 16-bit base and the unsigned 8-bit displacement. Because the addition is unsigned, the result is always a forward reference from the base of 0 to 255 bytes. The base register is either DPTR or PC.

**EXAMPLE 2.6 SOME INDEXED INSTRUCTIONS**

**MOVC A,@A+PC;**   Move code byte relative to PC to the accumulator.
**JMP @A+DPTR;**   Jump relative to DPTR.

### 2.10.6 Operand Modifiers: @ and #

The 8051 assembly language uses two special symbols to distinguish operand types: the at symbol (@) and the number sign (#). The @ before an operand means that indirect addressing is being used; # before an operand means it is an immediate operand (a constant).

**EXAMPLE 2.7 USE OF OPERAND MODIFIERS**

**ADD A,R2;**   Add to the accumulator the contents of register 2 of the selected bank.
**ADD A,@R2;**   Add to the accumulator the contents of the RAM location whose address is in register 2 of the selected bank.
**ADD A, 05;**   Add to the accumulator the contents of RAM address 05.
**ADD A,#05;**   Add to the accumulator the number 05.

## 2.11 INSTRUCTION GROUPS

The 8051 has 111 instruction types: 49 one-byte, 45 two-byte, and 17 three-byte. Counting the variations within each type, there are 255 separate instructions; every hex code from 00 to FF corresponds to a valid instruction except A5, which is reserved for future use. The instructions fall into five groups: arithmetic, logic, data transfer, Boolean, and branching. A brief description of each group is given below. Refer to Appendix D for a complete description.

### 2.11.1 Arithmetic Operations

The 8051 has the usual operations of an 8-bit processor: add (ADD), add-with-carry (ADDC), subtract-with-borrow (SUBB), increment (INC), decrement (DEC), and decimal-adjust-accumulator (DA). It also has two operations not typical of 8-bit micros: multiply (MUL AB) and divide (DIV AB).

MUL AB multiplies the contents of the accumultor by the contents of the B register as unsigned 8-bit integers to give a 16-bit result. The low-order 8 bits of the result will be in the accumulator and the high-order 8 bits will be in B. DIV AB causes the contents of the accumulator to be divided by the contents of B as unsigned integer numbers. The integer part of the result will be in the accumultor, and the remainder is left in B.

The addressing modes used by arithmetic instructions are direct, indirect, register, and immediate.

### 2.11.2 Logic Operations

The logic operations include *and* (ANL), *or* (ORL), *exclusive-or* (XRL), *clear* and *complement* (CLR and CPL), and *rotates* (RL, RLC, RR, RRC). Also included is the SWAP A instruction. SWAP A swaps nibbles within the accumulator. That is, it exchanges the lower 4 bits in the accumulator with the upper 4 bits. SWAP is useful for working with 4-bit quantities such as BCD numbers, which can be packed two to the byte.

Note that neither the arithmetic nor the logic group contains a compare instruction. The function of a compare instruction has been absorbed into a special branching instruction, as discussed in Sec. 2.11.5.

### 2.11.3 Data Transfers

The basic data transfer instruction is *move,* which has three forms: MOV, MOVC, and MOVX. Also included in this group are PUSH, POP, and XCH (exchange). All addressing modes are used in this group.

MOV instructions are used to reference internal RAM and SFR space. The two MOVC instructions are used to move bytes from program memory into the accumulator, as from a data table. MOVX instructions are used to reference external RAM.

### 2.11.4 Boolean Operations

This group of instructions is associated with the single-bit Boolean processor hardware of the 8051. The group includes *set* and *clear,* as well as *and, or,* and *complement* instructions. Also included are bit level move instructions and conditional *jumps,* which test bit values. In effect, this group makes up a miniature assembly language for the Boolean processor.

### 2.11.5 Branching Instructions

Included in this group are the subroutine calls and returns, as well as various conditional and unconditional jumps. The conditional jumps are relative to the first byte of the next instruction. Because the jump is given as a signed two's complement 8-bit number, the range is $-128$ to $+127$ bytes, thereby allowing forward and backward branching.

The jump instruction has three basic versions: the short jump (SJMP), the long jump (LJMP), and the absolute jump (AJMP). The *short jump* uses a relative offset as described above. The *long jump* (as well as the long call) uses a 16-bit address as part of the instruction, which makes the instruction 3 bytes long but enables it to reference any location in the 64K program memory space. The *absolute jump* (and the absolute call) uses an 11-bit address, which is split into 8 lower bits and 3 upper bits. The 3 upper bits are combined with a 5-bit operation specifier to make an 8-bit op-code. Thus, the entire instruction is only 2 bytes long. During execution, the 11 bits of the address are substituted for the lower 11 bits in the program counter (PC), which means that the location referenced must be within 2K bytes of the instruction following the AJMP (the upper 5 bits in PC remain the same).

An important instruction in this group is CJNE, which combines the functions of separate compare and jump instructions.

---

**EXAMPLE 2.8 USING CJNE**

CJNE A,07,03 will compare the contents of the accumulator with the contents of direct RAM address 07. If they are not equal, a jump is made to the instruction 3 bytes past the beginning of the next instruction. If they are equal, the next instruction, which could be an unconditional jump, is executed.

---

Also included in this group, by default, is NOP: a 1-byte instruction that does nothing, and takes one machine cycle to do it. One use for the NOP is in padding out delay loops to get exact times. Often a device attached to an output port pin will require a pulse of some minimum duration. NOPs can be used to determine the pulse width.

## 2.12 SINGLE-BIT BOOLEAN PROCESSOR

The Boolean processor can be thought of as a built-in bit-level coprocessor complete with its own set of instructions: the Boolean group. (It is not a true coprocessor; it doesn't run separately.) All the port lines, as well as 128 bits in RAM and many bits in the SFR registers, have bit addresses. Such a hardware and software combination makes the 8051 well suited for control applications that have many on/off kinds of inputs and outputs, such as switches, lamps, relays, stepper motor drives, and the like.

### 2.12.1 Carry: The Boolean Accumulator

The carry bit (bit 7 of the PSW) is the equivalent of an accumulator for the Boolean processor. The symbolic name CY is used to designate the carry bit when referring to it as a bit address; the symbol C is used in register-specific instructions that reference the carry bit. For example, CLR C is a 1-byte register-specific instruction that clears the carry bit to 0; CLR CY is a 2-byte instruction that clears the carry bit by referencing its address in the SFR space. The assembler will translate the symbol CY to the number D7H, the address of the carry bit. The instruction CLR PSW.7 is equivalent to CLR CY; it will generate the same code. CLR C generates different code but has the same effect.

Bits from the SFR registers, from internal RAM, and from I/O ports can be read into CY. Operations such as AND and OR can be performed on CY and the result written back to a bit address. Program branching can be conditional on the state of CY or any other addressable bit. Extensive bit manipulation can be done without having to use extraneous code to mask off bits to extract them from bytes, as is done in processors such as the 8085. Also, operations and testing can be done directly on bits without first moving them to CY.

### 2.12.2 An Application Example

The following example, taken from Intel application note AP-70, illustrates the advantage of a Boolean processor.

## EXAMPLE 2.9 A BOOLEAN APPLICATION

Figure 2.15 shows a logic circuit implemented with both gates and relays. The Boolean equation for the circuit is

$$Q = (U \cdot (V + W)) + (X \cdot \overline{Y}) + \overline{Z}$$

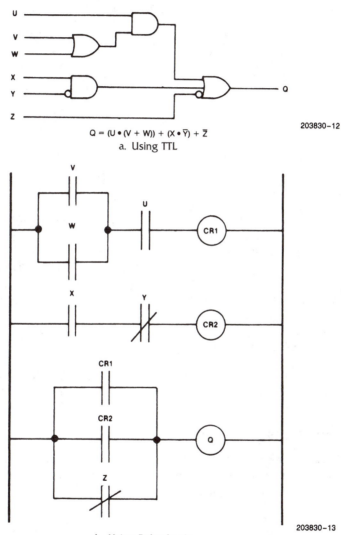

$Q = (U \bullet (V + W)) + (X \bullet \overline{Y}) + \overline{Z}$

a. Using TTL

203830-12

b. Using Relay Logic

203830-13

**Figure 2.15** Hardware implementations of Boolean functions. (*Source:* Reprinted by permission of Intel Corporation, Copyright © Intel Corporation 1987.)

The 8051 program segment to implement the equation follows. Note the elegance of the straight-line code made possible by the use of Boolean operations. It is left as an exercise for the student to implement the same equation in 8085 code; the comparison should be enlightening.

```
                          *
                          *
                          *

     MOV    C,V          ;GET V INPUT
     ORL    C,W          ;OUTPUT OF OR GATE
     ANL    C,U          ;OUTPUT OF TOP AND GATE
     MOV    F0,C         ;SAVE INTERMEDIATE STATE
     MOV    C,X          ;GET X INPUT
     ANL    C,Y          ;OUTPUT OF BOTTOM AND GATE
     ORL    C,F0         ;INCLUDE SAVED STATE
     ORL    C,Z          ;INCLUDE Z INPUT
     MOV    Q,C          ;OUTPUT COMPUTED RESULT

                          *
                          *
                          *
```

## 2.13 EXAMPLE PROGRAMS

Space does not allow for an exhaustive examination of every 8051 instruction. In this section we look at a few short applications to get a feel for 8051 programming. As you read the code, it will be useful to refer to the appropriate sections in this chapter.

Note the use of predefined bit and byte names, such as TCON, in the example programs. They are a common feature of 8051 assemblers. Such names, along with the op-codes and directives, are called *reserved symbols* and should not be used except for their intended purposes.

### 2.13.1 An Easy Move

Some of the things to notice in the program of Example 2.10 are the selection of the register bank, the initialization of the stack pointer, and the use of indexed addressing with the DPTR register.

### EXAMPLE 2.10 DATA MOVES

This small procedure is programmed in 8051 code that reads 10 bytes from a data table in program memory and stores each byte in a RAM location. After each byte is read, a subroutine is called, which, presumably, uses the data byte.

```
                          *
                          *
                          *

     ;              SELECT REGISTER BANK 1

          SETB    C           ; SET CARRY TO 1 FOR
          MOV     RS1,C       ; LSB OF BANK SELECT
          CLR     C           ; CLEAR CARRY TO O FOR
          MOV     RS0,C       ; MSB OF BANK SELECT
```

```
;                       MOVE STACK POINTER

        MOV     SP,#2FH         ; START STACK AT 30H

;                       INITIALIZE LOOP

        MOV     R0,00           ; INITIAL VALUE OF INDEX
        MOV     R2,0AH          ; LOOP COUNT = 10
        MOV     R1,20H          ; INITIAL RAM ADDR FOR SAVING
        MOV     DPTR,#0200H     ; BASE ADDRESS OF TABLE IN ROM

;                       EXECUTE LOOP

LOOP:   MOV     A,R0            ; MOVE INDEX VAL TO ACCUMULATOR
        MOVC    A,@A+DPTR       ; GET BYTE FROM TABLE
        MOV     @R1,A           ; STORE BYTE IN RAM
        ACALL   SUB             ; CALL ROUTINE THAT USES BYTE
        INC     R0              ; INCREMENT THE INDEX
        INC     R1              ; INCREMENT RAM ADDRESS
        DJNZ    R2,LOOP         ; DEC COUNT, JUMP IF NOT 0

                        <REST OF PROGRAM>

                                *
                                *
                                *
```

## 2.13.2 Jump Table

Example 3.6 shows the use of a jump-table in 8051 code. Note the use of base-displacement addressing.

### EXAMPLE 2.11 JUMP TABLE

The DPTR register is initialized to the beginning address of the table. The variable NUMBER is used to form the displacement. NUMBER is the sequential integer number of the program to which we will jump; it can have the value 1, 2, 3, or 4. Note that no check is made for values outside that range. The displacement, held in A, must be adjusted prior to the JMP @A+DPTR instruction. Because LJMP is 3 bytes long, NUMBER must be converted to a multiple of 3 (0, 3, 6, 9) before it is added to DPTR.

```
JTABLE:     MOV     DPTR,#TABLE ; GET TABLE ADDRESS AS BASE
            MOV     A,NUMBER      GET PROGRAM NUMBER
            RL      A           ; MULTIPLY NUMBER BY 2
            ADD     A,NUMBER    ; GET 3 TIMES NUMBER
            JMP     @A+DPTR     ; HOP TO IT


TABLE:      LJMP    PROG1       ; NOTE THAT
            LJMP    PROG2       ; LONG JUMPS
            LJMP    PROG3       ; ARE 3 BYTES
            LJMP    PROG4       ; IN LENGTH
```

## 2.13.3 ASCII to Hex Conversion

The routines shown in Example 2.12 were adapted from Silicon System's K224DEMO package, and are good examples of typical 8051 code. The code is given in the following program.

### EXAMPLE 2.12 ASCII TO HEX

```
; ROUTINE TO CONVERT 2 ASCII CHARACTERS TO HEX
;     R1 POINTS TO FIRST ASCII CHARACTER
;     HEX NUMBER RETURNED IN ACCUMULATOR
;     ERROR CAUSES RETURN WITH CARRY SET

A2HEX:
        MOV     R7,#2           ; INITIALIZE LOOP COUNT
CNV_1:
        MOV     A,@R1           ; GET ASCII CHARACTER
        LCALL   VAL_HEX         ; IS IT 0-9 OR A-F ?
        JC      INVHEX          ; IF NOT, JUMP TO RETURN
        CJNE    R7,#2,CNV_2     ; 2ND CHAR? JUMP IF YES
        SWAP    A               ; SWAP NIBBLES IN ACC
        MOV     R6,A            ; SAVE MSD
        INC     R1              ; POINT TO NEXT CHAR
        DJNZ    R7,CNV_1        ; FINISHED? JUMP IF NO
CNV_2:
        ORL     A,R6            ; COMBINE HALVES
        CLR     C               ; NO ERROR, CLEAR CARRY
INVHEX:
        RET                     ; GO HOME


; SUBROUTINE TO CHECK FOR VALID ASCII HEX DIGIT
;     RETURNS LOWER 4 BITS OF VALID DIGIT
;     SETS CARRY HIGH IF INVALID

VAL_HEX:
        CLR     C               ; CLEAR CARRY INITIALLY
        SUBB    A,#30H          ; REDUCES TO HEX IF 0-9
        JC      BADHEX          ; JUMP IF IT'S BELOW '0'
        SUBB    A,#0AH          ; BIGGER THAN '9' ?
        JC      VH_2            ; JUMP IF NO
        SUBB    A,#27H          ; BETWEEN '9' AND 'A' ?
        JC      BADHEX          ; JUMP TO ERROR IF YES
        SUBB    A,#06H          ; TEST FOR A-F
        JC      VH_1            ; JUMP IF IN RANGE
        SETB    C               ; SET ERROR FLAG
        SJMP    BADHEX          ; JUMP TO RETURN
VH_1:
        ADD     A,#06H          ; RENORMALIZE HEX
VH_2:
        ADD     A,#0AH          ; FIX UP 0-9
        CLR     C               ; SET NO-ERROR FLAG
BADHEX:
        RET                     ; GO BACK
```

Some features to note in the above code are:

1. The use of registers to hold the working variables and the loop counter. Register instructions are short and quick.

2. The use of carry (C) as an error flag.

3. The use of SWAP to get at the lower 4 bits of a character.

4. The use of CJNE to combine compare and conditional jump.

5. The use of DJNZ to implement a loop.

6. Because subtract-with-borrow (SUBB) is the only subtraction the 8051 has, the carry bit must be cleared to effect a simple subtraction.

7. The use of @ and # operand modifiers.

## 2.14 SYSTEM EXAMPLES: EXTERNAL ROM, RAM, I/O

Because so much is included on the chip, microcontrollers typically do not require much in the way of peripheral chip support. More often, a device such as the 8051 is interfaced directly to the input and output devices it is controlling. However, in order to illustrate the use of the 8051 two simple examples are included here.

### EXAMPLE 2.13 EXTERNAL PROGRAM MEMORY

Figure 2.16 shows the 8051 interfaced to a 2732 EPROM. Note the use of the ALE and PSEN pins, as well as the use of port 0 pins as data lines. Also note the use of port 2 pins as high-order address lines.

### EXAMPLE 2.14 EXTERNAL RAM AND I/O EXPANSION

Figure 2.17 shows the 8051 interfaced to an 8155 peripheral device. Note the use of $\overline{RD}$ and $\overline{WR}$ as strobes to the RAM data store. Also note the use of P2.0 to control the $1O/\overline{M}$ pin on the 8155.

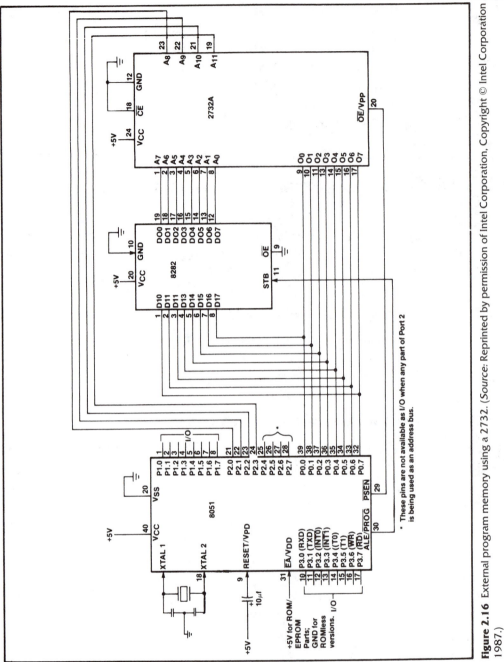

**Figure 2.16** External program memory using a 2732. (*Source:* Reprinted by permission of Intel Corporation, Copyright © Intel Corporation 1987.)

52

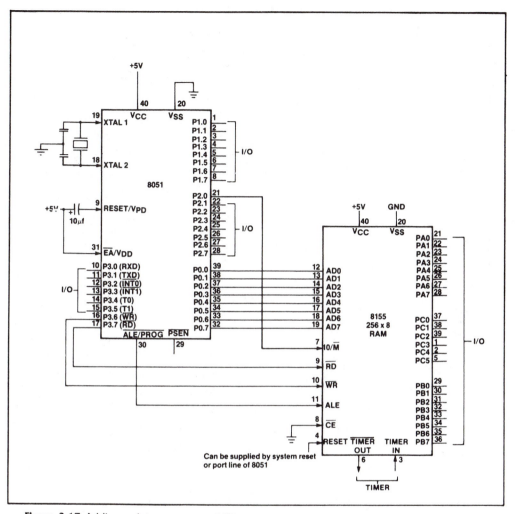

**Figure 2.17** Adding a data memory and I/O expander. (*Source:* Reprinted by permission of Intel Corporation, Copyright © Intel Corporation 1987.)

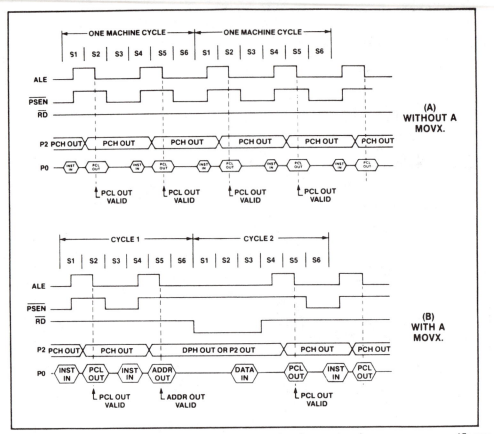

**Figure 2.18** Bus cycles in MCS©-51 devices executing from external program memory. (*Source:* Reprinted by permission of Intel Corporation, Copyright © Intel Corporation 1987.)

## 2.15 SUMMARY

Microcontrollers are devices that combine memory, I/O, and other functions with a CPU to form a sophisticated component level device for use in embedded control applications. The 8051, as an example of a microcontroller, has been examined in terms of its hardware and software. Key features include

- On-board RAM and ROM
- 32 I/O lines
- Hardware timer/counters
- On-board USART
- Sophisticated vectored interrupt system
- Boolean processor
- RAM resident control registers

# 8051 INSTRUCTION SET SUMMARY

Interrupt Response Time: Refer to Hardware Description Chapter.

## Instructions that Affect Flag Settings(1)

| Instruction | Flag | | | Instruction | Flag | | |
|---|---|---|---|---|---|---|---|
| | C | OV | AC | | C | OV | AC |
| ADD | X | X | X | CLR C | O | | |
| ADDC | X | X | X | CPL C | X | | |
| SUBB | X | X | X | ANL C,bit | X | | |
| MUL | O | X | | ANL C,/bit | X | | |
| DIV | O | X | | ORL C,bit | X | | |
| DA | X | | | ORL C,bit | X | | |
| RRC | X | | | MOV C,bit | X | | |
| RLC | X | | | CJNE | X | | |
| SETB C | 1 | | | | | | |

(1)Note that operations on SFR byte address 208 or bit addresses 209-215 (i.e., the PSW or bits in the PSW) will also affect flag settings.

### Note on instruction set and addressing modes:

Rn — Register R7–R0 of the currently selected Register Bank.

direct — 8-bit internal data location's address. This could be an Internal Data RAM location (0–127) or a SFR [i.e., I/O port, control register, status register, etc. (128–255)].

@Ri — 8-bit internal data RAM location (0–255) addressed indirectly through register R1 or R0.

#data — 8-bit constant included in instruction.

#data 16 — 16-bit constant included in instruction.

addr 16 — 16-bit destination address. Used by LCALL & LJMP. A branch can be anywhere within the 64K-byte Program Memory address space.

addr 11 — 11-bit destination address. Used by ACALL & AJMP. The branch will be within the same 2K-byte page of program memory as the first byte of the following instruction.

rel — Signed (two's complement) 8-bit offset byte. Used by SJMP and all conditional jumps. Range is −128 to +127 bytes relative to first byte of the following instruction.

bit — Direct Addressed bit in Internal Data RAM or Special Function Register.

* — New operation not provided by 8048AH/8049AH.

| Mnemonic | | Description | Byte | Oscillator Period |
|---|---|---|---|---|
| **ARITHMETIC OPERATIONS** | | | | |
| ADD | A,Rn | Add register to Accumulator | 1 | 12 |
| ADD | A,direct | Add direct byte to Accumulator | 2 | 12 |
| ADD | A,@Ri | Add indirect RAM to Accumulator | 1 | 12 |
| ADD | A,#data | Add immediate data to Accumulator | 2 | 12 |
| ADDC | A,Rn | Add register to Accumulator with Carry | 1 | 12 |
| ADDC | A,direct | Add direct byte to Accumulator with Carry | 2 | 12 |
| ADDC | A,@Ri | Add indirect RAM to Accumulator with Carry | 1 | 12 |
| ADDC | A,#data | Add immediate data to Acc with Carry | 2 | 12 |
| SUBB | A,Rn | Subtract Register from Acc with borrow | 1 | 12 |
| SUBB | A,direct | Subtract direct byte from Acc with borrow | 2 | 12 |
| SUBB | A,@Ri | Subtract indirect RAM from ACC with borrow | 1 | 12 |
| SUBB | A,#data | Subtract immediate data from Acc with borrow | 2 | 12 |
| INC | A | Increment Accumulator | 1 | 12 |
| INC | Rn | Increment register | 1 | 12 |
| INC | direct | Increment direct byte | 2 | 12 |
| INC | @Ri | Increment direct RAM | 1 | 12 |
| DEC | A | Decrement Accumulator | 1 | 12 |
| DEC | Rn | Decrement Register | 1 | 12 |
| DEC | direct | Decrement direct byte | 2 | 12 |
| DEC | @Ri | Decrement indirect RAM | 1 | 12 |

All mnemonics copyrighted © Intel Corporation 1980

| Mnemonic | | Description | Byte | Oscillator Period |
|---|---|---|---|---|
| **ARITHMETIC OPERATIONS** (Continued) | | | | |
| INC | DPTR | Increment Data Pointer | 1 | 24 |
| MUL | AB | Multiply A & B | 1 | 48 |
| DIV | AB | Divide A by B | 1 | 48 |
| DA | A | Decimal Adjust Accumulator | 1 | 12 |
| **LOGICAL OPERATIONS** | | | | |
| ANL | A,Rn | AND Register to Accumulator | 1 | 12 |
| ANL | A,direct | AND direct byte to Accumulator | 2 | 12 |
| ANL | A,@Ri | AND indirect RAM to Accumulator | 1 | 12 |
| ANL | A,#data | AND immediate data to Accumulator | 2 | 12 |
| ANL | direct,A | AND Accumulator to direct byte | 2 | 12 |
| ANL | direct,#data | AND immediate data to direct byte | 3 | 24 |
| ORL | A,Rn | OR register to Accumulator | 1 | 12 |
| ORL | A,direct | OR direct byte to Accumulator | 2 | 12 |
| ORL | A,@Ri | OR indirect RAM to Accumulator | 1 | 12 |
| ORL | A,#data | OR immediate data to Accumulator | 2 | 12 |
| ORL | direct,A | OR Accumulator to direct byte | 2 | 12 |
| ORL | direct,#data | OR immediate data to direct byte | 3 | 24 |
| XRL | A,Rn | Exclusive-OR register to Accumulator | 1 | 12 |
| XRL | A,direct | Exclusive-OR direct byte to Accumulator | 2 | 12 |
| XRL | A,@Ri | Exclusive-OR indirect RAM to Accumulator | 1 | 12 |
| XRL | A,#data | Exclusive-OR immediate data to Accumulator | 2 | 12 |
| XRL | direct,A | Exclusive-OR Accumulator to direct byte | 2 | 12 |
| XRL | direct,#data | Exclusive-OR immediate data to direct byte | 3 | 24 |
| CLR | A | Clear Accumulator | 1 | 12 |
| CPL | A | Complement Accumulator | 1 | 12 |

| Mnemonic | | Description | Byte | Oscillator Period |
|---|---|---|---|---|
| **LOGICAL OPERATIONS** (Continued) | | | | |
| RL | A | Rotate Accumulator Left | 1 | 12 |
| RLC | A | Rotate Accumulator Left through the Carry | 1 | 12 |
| RR | A | Rotate Accumulator Right | 1 | 12 |
| RRC | A | Rotate Accumulator Right through the Carry | 1 | 12 |
| SWAP | A | Swap nibbles within the Accumulator | 1 | 12 |
| **DATA TRANSFER** | | | | |
| MOV | A,Rn | Move register to Accumulator | 1 | 12 |
| MOV | A,direct | Move direct byte to Accumulator | 2 | 12 |
| MOV | A,@Ri | Move indirect RAM to Accumulator | 1 | 12 |
| MOV | A,#data | Move immediate data to Accumulator | 2 | 12 |
| MOV | Rn,A | Move Accumulator to register | 1 | 12 |
| MOV | Rn,direct | Move direct byte to register | 2 | 24 |
| MOV | Rn,#data | Move immediate data to register | 2 | 12 |
| MOV | direct,A | Move Accumulator to direct byte | 2 | 12 |
| MOV | direct,Rn | Move register to direct byte | 2 | 24 |
| MOV | direct,direct | Move direct byte to direct | 3 | 24 |
| MOV | direct,@Ri | Move indirect RAM to direct byte | 2 | 24 |
| MOV | direct,#data | Move immediate data to direct byte | 3 | 24 |
| MOV | @Ri,A | Move Accumulator to indirect RAM | 1 | 12 |

All mnemonics copyrighted © Intel Corporation 1980

Chap. 2: The 8051 Single-Chip Microcontroller

| Mnemonic | | Description | Byte | Oscillator Period |
|---|---|---|---|---|
| **DATA TRANSFER** (Continued) | | | | |
| MOV | @Ri,direct | Move direct byte to indirect RAM | 2 | 24 |
| MOV | @Ri, # data | Move immediate data to indirect RAM | 2 | 12 |
| MOV | DPTR, # data16 | Load Data Pointer with a 16-bit constant | 3 | 24 |
| MOVC | A,@A + DPTR | Move Code byte relative to DPTR to Acc | 1 | 24 |
| MOVC | A,@A + PC | Move Code byte relative to PC to Acc | 1 | 24 |
| MOVX | A,@Ri | Move External RAM (8-bit addr) to Acc | 1 | 24 |
| MOVX | A,@DPTR | Move External RAM (16-bit addr) to Acc | 1 | 24 |
| MOVX | @Ri,A | Move Acc to External RAM (8-bit addr) | 1 | 24 |
| MOVX | @DPTR,A | Move Acc to External RAM (16-bit addr) | 1 | 24 |
| PUSH | direct | Push direct byte onto stack | 2 | 24 |
| POP | direct | Pop direct byte from stack | 2 | 24 |
| XCH | A,Rn | Exchange register with Accumulator | 1 | 12 |
| XCH | A,direct | Exchange direct byte with Accumulator | 2 | 12 |
| XCH | A,@Ri | Exchange indirect RAM with Accumulator | 1 | 12 |
| XCHD | A,@Ri | Exchange low-order Digit indirect RAM with Acc | 1 | 12 |

| Mnemonic | | Description | Byte | Oscillator Period |
|---|---|---|---|---|
| **BOOLEAN VARIABLE MANIPULATION** | | | | |
| CLR | C | Clear Carry | 1 | 12 |
| CLR | bit | Clear direct bit | 2 | 12 |
| SETB | C | Set Carry | 1 | 12 |
| SETB | bit | Set direct bit | 2 | 12 |
| CPL | C | Complement Carry | 1 | 12 |
| CPL | bit | Complement direct bit | 2 | 12 |
| ANL | C,bit | AND direct bit to CARRY | 2 | 24 |
| ANL | C,/bit | AND complement of direct bit to Carry | 2 | 24 |
| ORL | C,bit | OR direct bit to Carry | 2 | 24 |
| ORL | C,/bit | OR complement of direct bit to Carry | 2 | 24 |
| MOV | C,bit | Move direct bit to Carry | 2 | 12 |
| MOV | bit,C | Move Carry to direct bit | 2 | 24 |
| JC | rel | Jump if Carry is set | 2 | 24 |
| JNC | rel | Jump if Carry not set | 2 | 24 |
| JB | bit,rel | Jump if direct Bit is set | 3 | 24 |
| JNB | bit,rel | Jump if direct Bit is Not set | 3 | 24 |
| JBC | bit,rel | Jump if direct Bit is set & clear bit | 3 | 24 |
| **PROGRAM BRANCHING** | | | | |
| ACALL | addr11 | Absolute Subroutine Call | 2 | 24 |
| LCALL | addr16 | Long Subroutine Call | 3 | 24 |
| RET | | Return from Subroutine | 1 | 24 |
| RETI | | Return from interrupt | 1 | 24 |
| AJMP | addr11 | Absolute Jump | 2 | 24 |
| LJMP | addr16 | Long Jump | 3 | 24 |
| SJMP | rel | Short Jump (relative addr) | 2 | 24 |

All mnemonics copyrighted ©Intel Corporation 1980

| Mnemonic | | Description | Byte | Oscillator Period |
|---|---|---|---|---|
| **PROGRAM BRANCHING** (Continued) | | | | |
| JMP | @A + DPTR | Jump indirect relative to the DPTR | 1 | 24 |
| JZ | rel | Jump if Accumulator is Zero | 2 | 24 |
| JNZ | rel | Jump if Accumulator is Not Zero | 2 | 24 |
| CJNE | A,direct,rel | Compare direct byte to Acc and Jump if Not Equal | 3 | 24 |
| CJNE | A,#data,rel | Compare immediate to Acc and Jump if Not Equal | 3 | 24 |

| Mnemonic | | Description | Byte | Oscillator Period |
|---|---|---|---|---|
| **PROGRAM BRANCHING** (Continued) | | | | |
| CJNE | Rn,#data,rel | Compare immediate to register and Jump if Not Equal | 3 | 24 |
| CJNE | @Ri,#data,rel | Compare immediate to indirect and Jump if Not Equal | 3 | 24 |
| DJNZ | Rn,rel | Decrement register and Jump if Not Zero | 2 | 24 |
| DJNZ | direct,rel | Decrement direct byte and Jump if Not Zero | 3 | 24 |
| NOP | | No Operation | 1 | 12 |

All mnemonics copyrighted © Intel Corporation 1980

*Source:* Reprinted by permission of Intel Corporation, copyright © Intel Corporation.

## PREDEFINED BYTE ADDRESSES

| Symbol | Hex Address | Meaning |
|---|---|---|
| ACC | E0 | ACCUMULATOR |
| B | F0 | MULTIPLICATION REGISTER |
| DPH | 83 | DATA POINTER <high byte> |
| DPL | 82 | DATA POINTER <low byte> |
| IE | A8 | INTERRUPT ENABLE |
| IP | B8 | INTERRUPT PRIORITY |
| P0 | 80 | PORT 0 |
| P1 | 90 | PORT 1 |
| P2 | A0 | PORT 2 |
| P3 | B0 | PORT 3 |
| PSW | D0 | PROGRAM STATUS WORD |
| SBUF | 99 | SERIAL PORT BUFFER |
| SCON | 98 | SERIAL PORT CONTROL |
| SP | 81 | STACK POINTER |
| TCON | 88 | TIMER CONTROL |
| TH0 | 8C | TIMER 0 <high byte> |
| TH1 | 8D | TIMER 1 <high byte> |
| TL0 | 8A | TIMER 0 <low byte> |
| TL1 | 8B | TIMER 1 <low byte> |
| TMOD | 89 | TIMER MODE |

## PREDEFINED CODE ADDRESSES

| Symbol | Hex Address | Meaning |
|---|---|---|
| RESET | 00 | Power Up (Reset) |
| EXTI0 | 03 | External Interrupt 0 |
| TIMER0 | 0B | Timer 0 Interrupt |
| EXTI1 | 13 | External Interrupt 1 |
| TIMER1 | 1B | Timer 1 Interrupt |
| SINT | 23 | Serial Port Interrupt |

## PREDEFINED BIT ADDRESSES

| Sym. | Position | Hex | Meaning |
|---|---|---|---|
| CY | PSW.7 | D7 | CARRY FLAG |
| AC | PSW.6 | D6 | AUXILIARY CARRY FLAG |
| F0 | PSW.5 | D5 | FLAG 0 |
| RS1 | PSW.4 | D4 | REGISTER BANK SELECT BIT 1 |
| RS0 | PSW.3 | D3 | REGISTER BANK SELECT BIT 0 |
| OV | PSW.2 | D2 | OVERFLOW FLAG |
| P | PSW.0 | D0 | PARITY FLAG |
| TF1 | TCON.7 | 8F | TIMER 1 OVERFLOW FLAG |
| TR1 | TCON.6 | 8E | TIMER 1 RUN CONTROL BIT |
| TF0 | TCON.5 | 8D | TIMER 0 OVERFLOW FLAG |
| TR0 | TCON.4 | 8C | TIMER 0 RUN CONTROL BIT |
| IE1 | TCON.3 | 8B | INTERRUPT 1 EDGE FLAG |
| IT1 | TCON.2 | 8A | INTERRUPT 1 TYPE CONTROL BIT |
| IE0 | TCON.1 | 89 | INTERRUPT 0 EDGE FLAG |
| IT0 | TCON.0 | 88 | INTERRUPT 0 TYPE CONTROL BIT |
| SM0 | SCON.7 | 9F | SERIAL MODE CONTROL BIT 0 |
| SM1 | SCON.6 | 9E | SERIAL MODE CONTROL BIT 1 |
| SM2 | SCON.5 | 9D | SERIAL MODE CONTROL BIT 2 |
| REN | SCON.4 | 9C | RECEIVE ENABLE |
| TB8 | SCON.3 | 9B | TRANSMIT BIT 8 |
| RB8 | SCON.2 | 9A | RECEIVE BIT 8 |
| TI | SCON.1 | 99 | TRANSMIT INTERRUPT FLAG |
| RI | SCON.0 | 98 | RECEIVE INTERRUPT FLAG |
| EA | IE.7 | AF | ENABLE ALL INTERRUPTS |
| ES | IE.4 | AC | ENABLE SERIAL PORT INTERRUPT |
| ET1 | IE.3 | AB | ENABLE TIMER 1 INTERRUPT |
| EX1 | IE.2 | AA | ENABLE EXTERNAL INTERRUPT 1 |
| ET0 | IE.1 | A9 | ENABLE TIMER 0 INTERRUPT |
| EX0 | IE.0 | A8 | ENABLE EXTERNAL INTERRUPT 0 |
| PS | IP.4 | BC | PRIORITY OF SERIAL PORT INTERRUPT |
| PT1 | IP.3 | BB | PRIORITY OF TIMER 1 INTERRUPT |
| PX1 | IP.2 | BA | PRIORITY OF EXTERNAL INTERRUPT 1 |
| PT0 | IP.1 | B9 | PRIORITY OF TIMER 0 |
| PX0 | IP.0 | B8 | PRIORITY OF EXTERNAL INTERRUPT 0 |
| RD | P3.7 | B7 | READ DATA FOR EXTERNAL MEMORY |
| WR | P3.6 | B6 | WRITE DATA FOR EXTERNAL MEMORY |
| T1 | P3.5 | B5 | TIMER/COUNTER 1 EXTERNAL FLAG |
| T0 | P3.4 | B4 | TIMER/COUNTER 0 EXTERNAL FLAG |
| INT1 | P3.3 | B3 | INTERRUPT 1 INPUT PIN |
| INT0 | P3.2 | B2 | INTERRUPT 0 INPUT PIN |
| TXD | P3.1 | B1 | SERIAL PORT TRANSMIT PIN |
| RXD | P3.0 | B0 | SERIAL PORT RECEIVE PIN |

# TABLE OF INSTRUCTION OPCODES IN HEXADECIMAL ORDER

| Hex Code | Number of Bytes | Mnemonic | Operands |
|---|---|---|---|
| 00 | 1 | NOP | |
| 01 | 2 | AJMP | *code addr* |
| 02 | 3 | LJMP | *code addr* |
| 03 | 1 | RR | A |
| 04 | 1 | INC | A |
| 05 | 2 | INC | *data addr* |
| 06 | 1 | INC | @R0 |
| 07 | 1 | INC | @R1 |
| 08 | 1 | INC | R0 |
| 09 | 1 | INC | R1 |
| 0A | 1 | INC | R2 |
| 0B | 1 | INC | R3 |
| 0C | 1 | INC | R4 |
| 0D | 1 | INC | R5 |
| 0E | 1 | INC | R6 |
| 0F | 1 | INC | R7 |
| 10 | 3 | JBC | *bit addr,code addr* |
| 11 | 2 | ACALL | *code addr* |
| 12 | 3 | LCALL | *code addr* |
| 13 | 1 | RRC | A |
| 14 | 1 | DEC | A |
| 15 | 2 | DEC | *data addr* |
| 16 | 1 | DEC | @R0 |
| 17 | 1 | DEC | @R1 |
| 18 | 1 | DEC | R0 |
| 19 | 1 | DEC | R1 |
| 1A | 1 | DEC | R2 |
| 1B | 1 | DEC | R3 |
| 1C | 1 | DEC | R4 |
| 1D | 1 | DEC | R5 |
| 1E | 1 | DEC | R6 |
| 1F | 1 | DEC | R7 |
| 20 | 3 | JB | *bit addr,code addr* |
| 21 | 2 | AJMP | *code addr* |
| 22 | 1 | RET | |
| 23 | 1 | RL | A |
| 24 | 2 | ADD | A,#*data* |
| 25 | 2 | ADD | A,*data addr* |
| 26 | 1 | ADD | A,@R0 |
| 27 | 1 | ADD | A,@R1 |
| 28 | 1 | ADD | A,R0 |
| 29 | 1 | ADD | A,R1 |
| 2A | 1 | ADD | A,R2 |

# TABLE OF INSTRUCTION OPCODES IN HEXADECIMAL ORDER (Cont'd.)

| Hex Code | Number of Bytes | Mnemonic | Operands |
|---|---|---|---|
| 2B | 1 | ADD | A,R3 |
| 2C | 1 | ADD | A,R4 |
| 2D | 1 | ADD | A,R5 |
| 2E | 1 | ADD | A,R6 |
| 2F | 1 | ADD | A,R7 |
| 30 | 3 | JNB | *bit addr,code addr* |
| 31 | 2 | ACALL | *code addr* |
| 32 | 1 | RETI | |
| 33 | 1 | RLC | A |
| 34 | 2 | ADDC | A,#*data* |
| 35 | 2 | ADDC | A,*data addr* |
| 36 | 1 | ADDC | A,@R0 |
| 37 | 1 | ADDC | A,@R1 |
| 38 | 1 | ADDC | A,R0 |
| 39 | 1 | ADDC | A,R1 |
| 3A | 1 | ADDC | A,R2 |
| 3B | 1 | ADDC | A,R3 |
| 3C | 1 | ADDC | A,R4 |
| 3D | 1 | ADDC | A,R5 |
| 3E | 1 | ADDC | A,R7 |
| 3F | 1 | ADDC | A,R7 |
| 40 | 2 | JC | *code addr* |
| 41 | 2 | AJMP | *code addr* |
| 42 | 2 | ORL | *data addr*,A |
| 43 | 3 | ORL | *data addr*,#*data* |
| 44 | 2 | ORL | A,#*data* |
| 45 | 2 | ORL | A,*data addr* |
| 46 | 1 | ORL | A,@R0 |
| 47 | 1 | ORL | A,@R1 |
| 48 | 1 | ORL | A,R0 |
| 49 | 1 | ORL | A,R1 |
| 4A | 1 | ORL | A,R2 |
| 4B | 1 | ORL | A,R3 |
| 4C | 1 | ORL | A,R4 |
| 4D | 1 | ORL | A,R5 |
| 4E | 1 | ORL | A,R6 |
| 4F | 1 | ORL | A,R7 |
| 50 | 2 | JNC | *code addr* |
| 51 | 2 | ACALL | *code addr* |
| 52 | 2 | ANL | *data addr*,A |
| 53 | 3 | ANL | *data addr*,#*data* |
| 54 | 2 | ANL | A,#*data* |
| 55 | 2 | ANL | A,*data addr* |

Chap. 2: The 8051 Single-Chip Microcontroller

| Hex Code | Number of Bytes | Mnemonic | Operands |
|---|---|---|---|
| 56 | 1 | ANL | A,@R0 |
| 57 | 1 | ANL | A,@R1 |
| 58 | 1 | ANL | A,R0 |
| 59 | 1 | ANL | A,R1 |
| 5A | 1 | ANL | A,R2 |
| 5B | 1 | ANL | A,R3 |
| 5C | 1 | ANL | A,R4 |
| 5D | 1 | ANL | A,R5 |
| 5E | 1 | ANL | A,R6 |
| 5F | 1 | ANL | A,R7 |
| 60 | 2 | JZ | code addr |
| 61 | 2 | AJMP | code addr |
| 62 | 2 | XRL | data addr,A |
| 63 | 3 | XRL | data addr,#data |
| 64 | 2 | XRL | A,#data |
| 65 | 2 | XRL | A,data addr |
| 66 | 1 | XRL | A,@R0 |
| 67 | 1 | XRL | A,@R1 |
| 68 | 1 | XRL | A,R0 |
| 69 | 1 | XRL | A,R1 |
| 6A | 1 | XRL | A,R2 |
| 6B | 1 | XRL | A,R3 |
| 6C | 1 | XRL | A,R4 |
| 6D | 1 | XRL | A,R5 |
| 6E | 1 | XRL | A,R6 |
| 6F | 1 | XRL | A,R7 |
| 70 | 2 | JNZ | code addr |
| 71 | 2 | ACALL | code addr |
| 72 | 2 | ORL | C,bit addr |
| 73 | 1 | JMP | @A+DPTR |
| 74 | 2 | MOV | A,#data |
| 75 | 3 | MOV | data addr,#data |
| 76 | 2 | MOV | @R0,#data |
| 77 | 2 | MOV | @R1,#data |
| 78 | 2 | MOV | R0,#data |
| 79 | 2 | MOV | R1,#data |
| 7A | 2 | MOV | R2,#data |
| 7B | 2 | MOV | R3,#data |
| 7C | 2 | MOV | R4,#data |
| 7D | 2 | MOV | R5,#data |
| 7E | 2 | MOV | R6,#data |
| 7F | 2 | MOV | R7,#data |
| 80 | 2 | SJMP | code addr |

| Hex Code | Number of Bytes | Mnemonic | Operands |
|---|---|---|---|
| 81 | 2 | AJMP | code addr |
| 82 | 2 | ANL | C,bit addr |
| 83 | 1 | MOVC | A,@A+PC |
| 84 | 1 | DIV | AB |
| 85 | 3 | MOV | data addr,data addr |
| 86 | 2 | MOV | data addr,@R0 |
| 87 | 2 | MOV | data addr,@R1 |
| 88 | 2 | MOV | data addr,R0 |
| 89 | 2 | MOV | data addr,R1 |
| 8A | 2 | MOV | data addr,R2 |
| 8B | 2 | MOV | data addr,R3 |
| 8C | 2 | MOV | data addr,R4 |
| 8D | 2 | MOV | data addr,R5 |
| 8E | 2 | MOV | data addr,R6 |
| 8F | 2 | MOV | data addr,R7 |
| 90 | 3 | MOV | DPTR,#data |
| 91 | 2 | ACALL | code addr |
| 92 | 2 | MOV | bit addr,C |
| 93 | 1 | MOVC | A,@A+DPTR |
| 94 | 2 | SUBB | A,#data |
| 95 | 2 | SUBB | A,data addr |
| 96 | 1 | SUBB | A,@R0 |
| 97 | 1 | SUBB | A,@R1 |
| 98 | 1 | SUBB | A,R0 |
| 99 | 1 | SUBB | A,R1 |
| 9A | 1 | SUBB | A,R2 |
| 9B | 1 | SUBB | A,R3 |
| 9C | 1 | SUBB | A,R4 |
| 9D | 1 | SUBB | A,R5 |
| 9E | 1 | SUBB | A,R6 |
| 9F | 1 | SUBB | A,R7 |
| A0 | 2 | ORL | C,/bit addr |
| A1 | 2 | AJMP | code addr |
| A2 | 2 | MOV | C,bit addr |
| A3 | 1 | INC | DPTR |
| A4 | 1 | MUL | AB |
| A5 |  | reserved |  |
| A6 | 2 | MOV | @R0,data addr |
| A7 | 2 | MOV | @R1,data addr |
| A8 | 2 | MOV | R0,data addr |
| A9 | 2 | MOV | R1,data addr |
| AA | 2 | MOV | R2,data addr |
| AB | 2 | MOV | R3,data addr |

| Hex Code | Number of Bytes | Mnemonic | Operands |
|---|---|---|---|
| AC | 2 | MOV | R4,data addr |
| AD | 2 | MOV | R5,data addr |
| AE | 2 | MOV | R6,data addr |
| AF | 2 | MOV | R7,data addr |
| B0 | 2 | ANL | C,/bit addr |
| B1 | 2 | ACALL | code addr |
| B2 | 2 | CPL | bit addr |
| B3 | 1 | CPL | C |
| B4 | 3 | CJNE | A,#data,code addr |
| B5 | 3 | CJNE | A,data addr,code addr |
| B6 | 3 | CJNE | @R0,#data,code addr |
| B7 | 3 | CJNE | @R1,#data,code addr |
| B8 | 3 | CJNE | R0,#data,code addr |
| B9 | 3 | CJNE | R1,#data,code addr |
| BA | 3 | CJNE | R2,#data,code addr |
| BB | 3 | CJNE | R3,#data,code addr |
| BC | 3 | CJNE | R4,#data,code addr |
| BD | 3 | CJNE | R5,#data,code addr |
| BE | 3 | CJNE | R6,#data,code addr |
| BF | 3 | CJNE | R7,#data,code addr |
| C0 | 2 | PUSH | data addr |
| C1 | 2 | AJMP | code addr |
| C2 | 2 | CLR | bit addr |
| C3 | 1 | CLR | C |
| C4 | 1 | SWAP | A |
| C5 | 2 | XCH | A,data addr |
| C6 | 1 | XCH | A,@R0 |
| C7 | 1 | XCH | A,@R1 |
| C8 | 1 | XCH | A,R0 |
| C9 | 1 | XCH | A,R1 |
| CA | 1 | XCH | A,R2 |
| CB | 1 | XCH | A,R3 |
| CC | 1 | XCH | A,R4 |
| CD | 1 | XCH | A,R5 |
| CE | 1 | XCH | A,R6 |
| CF | 1 | XCH | A,R7 |
| D0 | 2 | POP | data addr |
| D1 | 2 | ACALL | code addr |
| D2 | 2 | SETB | bit addr |
| D3 | 1 | SETB | C |
| D4 | 1 | DA | A |
| D5 | 3 | DJNZ | data addr,code addr |
| D6 | 1 | XCHD | A,@R0 |

| Hex Code | Number of Bytes | Mnemonic | Operands |
|---|---|---|---|
| D7 | 1 | XCHD | A,@R1 |
| D8 | 2 | DJNZ | R0,code addr |
| D9 | 2 | DJNZ | R1,code addr |
| DA | 2 | DJNZ | R2,code addr |
| DB | 2 | DJNZ | R3,code addr |
| DC | 2 | DJNZ | R4,code addr |
| DD | 2 | DJNZ | R5,code addr |
| DE | 2 | DJNZ | R6,code addr |
| DF | 2 | DJNZ | R7,code addr |
| E0 | 1 | MOVX | A,@DPTR |
| E1 | 2 | AJMP | code addr |
| E2 | 1 | MOVX | A,@R0 |
| E3 | 1 | MOVX | A,@R1 |
| E4 | 1 | CLR | A |
| E5 | 2 | MOV | A,data addr |
| E6 | 1 | MOV | A,@R0 |
| E7 | 1 | MOV | A,@R1 |
| E8 | 1 | MOV | A,R0 |
| E9 | 1 | MOV | A,R1 |
| EA | 1 | MOV | A,R2 |
| EB | 1 | MOV | A,R3 |
| EC | 1 | MOV | A,R4 |
| ED | 1 | MOV | A,R5 |
| EE | 1 | MOV | A,R6 |
| EF | 1 | MOV | A,R7 |
| F0 | 1 | MOVX | @DPTR,A |
| F1 | 2 | ACALL | code addr |
| F2 | 1 | MOVX | @R0,A |
| F3 | 1 | MOVX | @R1,A |
| F4 | 1 | CPL | A |
| F5 | 2 | MOV | data addr,A |
| F6 | 1 | MOV | @R0,A |
| F7 | 1 | MOV | @R1,A |
| F8 | 1 | MOV | R0,A |
| F9 | 1 | MOV | R1,A |
| FA | 1 | MOV | R2,A |
| FB | 1 | MOV | R3,A |
| FC | 1 | MOV | R4,A |
| FD | 1 | MOV | R5,A |
| FE | 1 | MOV | R6,A |
| FF | 1 | MOV | R7,A |

## REFERENCES

*Intel Embedded Controller Handbook* (1987), Intel order number 210918-005.

*Intel Microcontroller Handbook* (1983), Intel order number 210918-001.

*Intel MCS-51 Macro Assembler User's Guide,* Intel order number 9800937-03.

*Intel MCS-51 Macro Assembler User's Guide for DOS Systems,* Intel order number 122753-001.

*The 8051: Programming, Interfacing, Applications* (Howard Boyet and Ron Katz, MTI Publications, Inc., 14 E. 8 St., New York, NY 10003).

## CHAPTER REVIEW

### Questions

1. Explain the term *embedded*.
2. How do open-loop and closed-loop applications differ?
3. How does the 8031 differ from the 8051?
4. How many clocks are in an 8051 machine cycle?
5. How many phases are in an 8051 machine cycle?
6. Explain how program memory and data memory are treated by the 8051.
7. What is the function of the EA pin?
8. What is in the SFR area?
9. What bits are in the PSW?
10. Can jump-on-zero be done without a zero flag bit? How?
11. What is in SP at power up?
12. What control register is associated with SBUF?
13. What control registers are associated with the timers?
14. Which port can drive the heaviest load?
15. Which port is used for address and data with external memory?
16. Which port has alternate functions for its pins?
17. How does the use of PSEN differ from the use of $\overline{RD}$ and $\overline{WR}$?
18. When can the contents of port 0 latches be lost?
19. When is a counter advanced from an external source?
20. Explain how mode 2 differs from mode 0 in a counter.
21. Explain a possible use for counter mode 3.
22. Explain double buffering.
23. Which serial port modes allow full duplex asynchronous operation with variable baud rate?

24. In what register is the SMOD bit found?

25. What does it mean to say an interrupt is pending?

26. Name the interrupt sources in the 8051.

27. Explain the two-tier interrupt priority system.

28. Assuming both interrupts are high level and occur simultaneously, which gets done first, RI or TF0?

29. Does the ordering of the within-level priorities make sense to you? Explain your answer.

30. Explain the special timing involved when an interrupt occurs during the execution of a RETI instruction. What might happen if the timing were otherwise?

31. What controls the state of the flag for an external level-activated interrupt?

32. Name an interrupt flag that is not automatically cleared when the hardware generates the call to the service routine.

33. How long must a reset level be applied to the 8051?

34. How does the 8051 distinguish between a byte address and a bit address?

35. Explain indexed addressing in the 8051.

36. What holds the index number?

37. How many instruction types does the 8051 have? How many do you think you would actually use?

38. Explain what a Boolean processor does. How does it differ from what a byte processor does?

39. Write a program that reads a byte from port 1 and uses it as an index into a jump table starting at address 02B0H.

40. Write a program that will read a byte from port 0 and, depending on which bit in the byte is a 1, jump to one of eight different locations. Use Boolean instructions.

## Problems

1. Select a crystal frequency to get a baud rate of 9600, assuming SMOD = 0 and TH1 contains 254. Will the 8051 run that fast?

2. Will a crystal frequency of 11.059 MHz allow the following baud rates: 600, 1200, 2400, 4800? What value or values are required to be in TH1?

3. Write the program suggested in Example 2.9.

## Chapter 3

# Interfacing:
# Hardware
# and Software

Upon completion of this chapter, you should be able to

1. Analyze a typical hardware/software interface
2. Write code to interface to various electromechanical devices
3. Write code to interface to various printers and displays
4. Explain interfaces such as Centronics, IEEE-488, RS-232C, and A/D
5. Write a simple interrupt I/O program

## 3.1 INTRODUCTION

When you stop to think about it, getting a microprocessor to do something for you in the real world is an amazing thing. First you have an idea of what you want done. Then you translate that idea into a program that is stored in the processor's memory. When the program executes, your idea acts itself out using the various I/O devices connected to the processor. It can get feedback through input sensors and change the flow of execution accordingly while it drives output devices to carry out its task. If we didn't know how it was done, we might be tempted to call it magic.

The place where information and control flow back and forth between the program and the real world is the interface. The exact meaning of the term *interface* depends on the context in which it is used, but usually it means more than just an I/O port. An interface can also include special hardware, such as an A/D converter, as well as special software, such as a subroutine to use the A/D (a driver routine). Also, information may have to be structured in a certain way to pass through the interface, as, for example, with a communications protocol such as RS-232. In this chapter we examine some typical interfacing issues, both hardware and software.

## 3.2 MECHANICAL SWITCHES

Mechanical switches are common input devices. They are discussed here in terms of hardware and software considerations.

### 3.2.1 Description and Nomenclature

A common mechanical switch construction is one or more pairs of contacts that can be *open* (not touching) or *closed* (touching). One contact of each pair is mounted on a movable piece called the *pole*. The pole can be in one of two positions, depending on whether or not the switch is *activated* (pressed). If a pole can close a contact pair in only one of the two positions, it is said to be *single throw*. If it can close a pair in both positions, it is called *double throw*. Also, *normally open* (N.O.) contacts close when a switch is activated, and *normally closed* (N.C.) contacts open when a switch is activated. Switches are drawn in their normal, deactivated (finger off the button) state.

Typical switch designations are SPST, for single-pole-single-throw, SPDT, for single-pole-double-throw, and DPDT, for double-pole-double-throw. When the number of poles exceeds two, a digit is used such as 4PDT for four-pole-double-throw. There are various mechanical means of actuation, such as push button, toggle, foot switch, and the like. A switch that is designed for use on machinery where some moving part of the machine will activate it is called a *limit switch*. Figure 3.1 shows the schematic symbols for several switches.

Mechanical switches also can be characterized as either *momentary* or *latched*. Many (but certainly not all) momentary switches are of the push-button type (see Fig. 3.2). A momentary switch stays activated only as long as you press it, whereas a latched switch (e.g., a toggle switch) will stay in the last position it was placed.

When choosing a switch, it must be "sized" for the application—that is, the contacts must be able to pass the required current and be able to interrupt the voltage levels present on opening without excessive arcing. An undersized switch will have a short life expectancy.

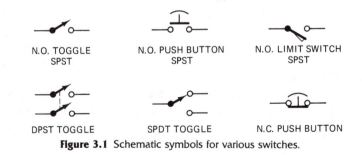

N.O. TOGGLE
SPST

N.O. PUSH BUTTON
SPST

N.O. LIMIT SWITCH
SPST

DPST TOGGLE

SPDT TOGGLE

N.C. PUSH BUTTON

**Figure 3.1** Schematic symbols for various switches.

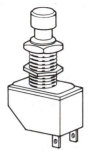

**Figure 3.2** Push-button switch.

## EXAMPLE 3.1 A DRY CIRCUIT

When a switch is used as an input to a microprocessor, it is usually a *dry circuit* application, meaning that the voltage and current levels are so low (microvolts and microamps) that they are not significant. Switch life can then be on the order of $10^4$ to $10^6$ activations, depending on mechanical quality.

A *rotary switch* (see Fig. 3.3) differs from those just described in that the rotating pole can have more than two throws. You could, for example, have a 4P10T rotary switch, but it would be called a four-pole-ten-position switch instead of a ten-throw switch. Rotaries are commonly used for selector switch types of applications. The contacts on rotary switches may be *break-before-make,* meaning that as the switch is rotated, the contact on the pole will leave one stationary contact completely (break) before it touches the next contact (make). Or else they can be *make-before-break,* meaning that as the switch is rotated, the contact on the pole will, for a brief time, short out two adjacent stationary contacts before coming to rest at the next position.

### 3.2.2 Contact Bounce and Debouncing

The moving parts of a switch have mass and springiness. What they don't have is a lot of resistance to their motion, or damping. Any springy mechanical system with low damping is going to exhibit oscillatory, or "bouncy," behavior. When a N.O. switch is activated,

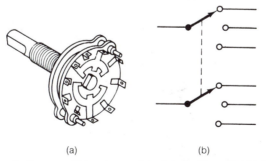

(a)                    (b)

**Figure 3.3** Two-pole-three-position rotary switch. Adapted from Patrick O'Conner, *Digital and Microprocessor Technology* © 1983, p. 145. Reprinted by permission of Prentice-Hall, Inc., Englewood Cliffs, N.J.

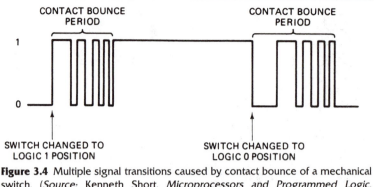

Figure 3.4 Multiple signal transitions caused by contact bounce of a mechanical switch. (*Source:* Kenneth Short, *Microprocessors and Programmed Logic,* © 1981, p. 332. Reprinted by permission of Prentice-Hall, Inc., Englewood Cliffs, N.J.)

the contacts will come together and bounce off each other several times before finally coming to rest in a closed position (see Fig. 3.4). Such behavior is called *contact bounce* or *switch bounce*. Contacts bounce over a period of milliseconds (5–25 msec), whereas microprocessors can execute an entire subroutine in a period of microseconds. A common example is a keyboard on a CRT terminal. Each key is actually a switch. If you press the letter E, you want a single E to appear on the screen. A bouncy switch can make the processor think the key was pressed several times, and several Es will appear. Obviously, mechanical switches must be *debounced*.

## EXAMPLE 3.2 HARDWARE DEBOUNCING

Debouncing may be done in hardware, as shown in Fig. 3.5. Note the use of *pull-up* resistors to provide logic level voltages to the switch circuit.

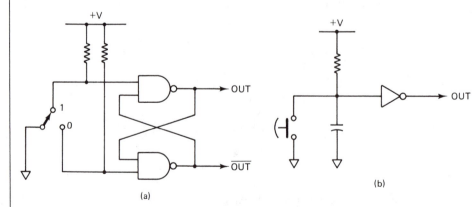

(a)

(b)

Figure 3.5 Debouncing.

Debouncing can also be done in software. Although a hardware solution may be practical for one or two switches, cost considerations often require a software solution for keypads and other applications using large numbers of switches. Note that, through use, switches tend to loose some of their springiness. The result is that the time it takes for bouncing to stop may *increase* as a switch gets older, and the debounce code that worked fine when the keypad was new may not work a year later unless you allowed for the change. Consult the switch manufacturer for data on worst case bounce time.

## EXAMPLE 3.3 SOFTWARE DEBOUNCING

The program here shows the subroutine PCNT, which counts the number of times a momentary push button is pressed. When PCNT is called, it waits in LOOP1 for the switch to be pressed. When a contact closure is detected, PCNT updates the count (stored in register R0) and calls the DELAY subroutine to kill time waiting for the switch to stop bouncing. After returning from DELAY, PCNT then waits in LOOP2 for you to take your finger off the switch before it returns to the calling routine. Note that LOOP2 is just as necessary as the debounce delay to prevent multiple counts from a single switch press.

```
                    ORG     100H           ; FOR THIS EXAMPLE

            SWITCH  DATA    P1             ; INPUT PORT FOR SWITCHES

COEO        PCNT:   PUSH    ACC            ; SAVE ACCUMULATOR
7590FF              MOV     SWITCH,#0FFH   ; MAKE PINS HIGH, INPUTS
                                           ; ARE CLOSURES TO GROUND
2090FD      LOOP1:  JB      SWITCH.0,LOOP1 ; LOOP UNTIL SWITCH CLOSES
                                           ; ON BIT 0 OF PORT 1

08                  INC     R0             ; ADD 1 TO COUNT
3113                ACALL   DELAY          ; WAIT TILL BOUNCING STOPS

3090FD      LOOP2:  JNB     SWITCH.0,LOOP2 ; LOOP UNTIL SWITCH OPENS
3113                ACALL   DELAY          ; CAN BOUNCE ON OPENING TOO

DOEO                POP     ACC            ; RESTORE ACCUMULATOR
22                  RET                    ; BYE
```

## 3.2.3 Switch Arrays and Encoders

If a system uses only a few switches, their status can be read directly through an I/O port. When many switches are used in an array, such as a 16-switch hex keypad or a 64-switch typewriter style keyboard, other methods are used to read them. One method is to use a hardware *encoder*. Some switches, such as the *thumbwheel* switch shown in Fig. 3.6, come with built-in encoders. More commonly a circuit is required.

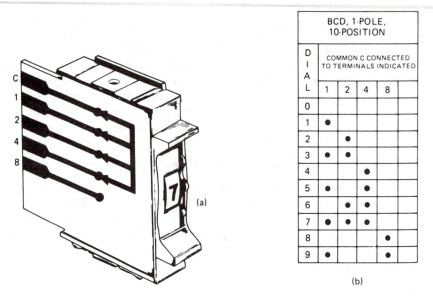

| BCD, 1-POLE, 10-POSITION | | | | | |
|---|---|---|---|---|---|
| D I A L | COMMON C CONNECTED TO TERMINALS INDICATED | | | | |
| | 1 | 2 | 4 | 8 | |
| 0 | | | | | |
| 1 | • | | | | |
| 2 | | • | | | |
| 3 | • | • | | | |
| 4 | | | • | | |
| 5 | • | | • | | |
| 6 | | • | • | | |
| 7 | • | • | • | | |
| 8 | | | | • | |
| 9 | • | | | • | |

(a)

(b)

**Figure 3.6** Thumbwheel switch. (a) Pictorial. (b) Truth table indicating which terminals are connected to common for each dial position. (*Source:* Adapted from Kenneth Short, *Microprocessors and Programmed Logic,* © 1981, p. 335. Reprinted by permission of Prentice-Hall, Inc., Englewood Cliffs, N.J.)

## EXAMPLE 3.4 ENCODING

In Fig. 3.7, 16 switches are shown connected to an encoder circuit. When any switch is pressed, the encoder will generate a 4-bit number corresponding to that particular switch. As soon as the 4-bit number becomes stable, a data-available (DAV) status bit will go true. When the switch is released, DAV goes false. Note that the encoder also debounces the switches, so DAV will go true only once for each key press.

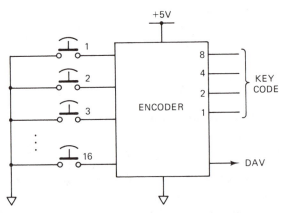

**Figure 3.7** Switches connected to an encoder.

Chap. 3: Interfacing: Hardware and Software

A problem occurs if more than one switch is pressed simultaneously. One solution is to have the encoder either ignore multiple key presses or output another status bit to indicate an error condition. A more useful approach is to have the encoder implement *rollover*. Rollover depends on the fact that even when two or more keys are pressed "simultaneously," in fact one switch will close before the other. In *two-key rollover*, the first contact closure is accepted and all subsequent closures are ignored until the first key is released. In *N-key rollover*, a second key press will be accepted before the first key is released. Although more complicated to implement, N-key rollover is preferred in applications such as terminal keyboards where rapid typing can easily cause multiple key presses.

### 3.2.4 Switch Matrix and Keyboards

For large numbers of switches, the direct connection shown in Fig. 3.7 is not practical because of the large number of connections required, one for each switch. The number of connections can be reduced by wiring the switches into rows and columns in an X-Y *matrix*, as shown in Fig. 3.8. The switches in a matrix are read by a *scanning process*.

In the typical matrix, the columns are connected to the logic 1 level by pull-up resistors. The rows are also held high but are driven low, one at a time, in a repeating cycle. Each time a row is driven low, the columns are read sequentially to see which, if any, are low. Because the switches are N.O., if no key is pressed all the columns will

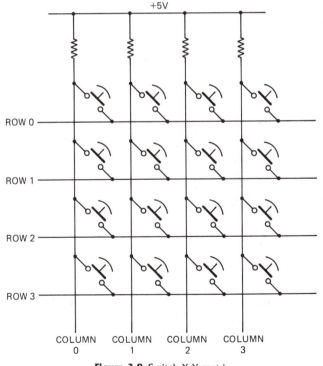

**Figure 3.8** Switch X-Y matrix.

remain high. If a key is pressed, a column will go low when the row that intersects that column at the closed switch is driven low. By knowing which row and column are low simultaneously, you can figure out which key was pressed. Note that the roles of the rows and columns can be reversed. Also note that the switches must be scanned rapidly compared to the speed of a key press, but that is not usually a problem.

A matrix can be scanned in software using I/O ports to drive the rows and read the columns. It can also be done in hardware with a chip designed for the purpose, such as the MM74C923 shown in Fig. 3.9. Some encoders just match a sequential binary number to each switch. When an encoder is built into a keyboard to form a separate device (a fully encoded keyboard), it is usually desirable that the encoder output a standard code for each labeled key. Thus, pressing the A key on a fully encoded ASCII keyboard will yield the hex number 41 at the output of the encoder together with a data-available signal. The encoder hardware may contain a combination of logic and ROM to generate the appropriate codes to match the keys. See Fig. 3.10.

A keyboard is not the only application for a switch matrix. Switch matrixes are often used to trace movement, such as pressure switches buried in a floor or fine wires embedded in a writing pad to follow the movement of a stylus.

## 3.3 SOLID STATE SWITCHES

If a microprocessor needs to read signals from an external transducer, such as a photo pickup, the voltage levels supplied by the device may not be compatible with the logic levels required at the I/O port. One solution is to use a transistor as a solid state N.O. switch.

### EXAMPLE 3.5 A TRANSISTOR SWITCH

Refer to Fig. 3.11. When sufficient voltage is applied to the base (or gate, if using a FET device), the transistor turns on and acts like a switch closed to ground. When the voltage is removed from the base, the transistor turns off and acts like an open switch.

Transistors are much faster than mechanical switches and can last almost indefinitely when used properly. There are, however, two problems to consider: switching and isolation.

### 3.3.1 Switching and Hysteresis

The *switching* problem comes about because for some range of base voltage the transistor can act like a linear amplifier. The result is that if the signal from the transducer changes from the on level to the off level slowly, the transistor may rapidly turn on and off many times as it passes through its linear region. The effect is similar to contact bounce. The solution is to introduce some *hysteresis*.

## Block Diagram

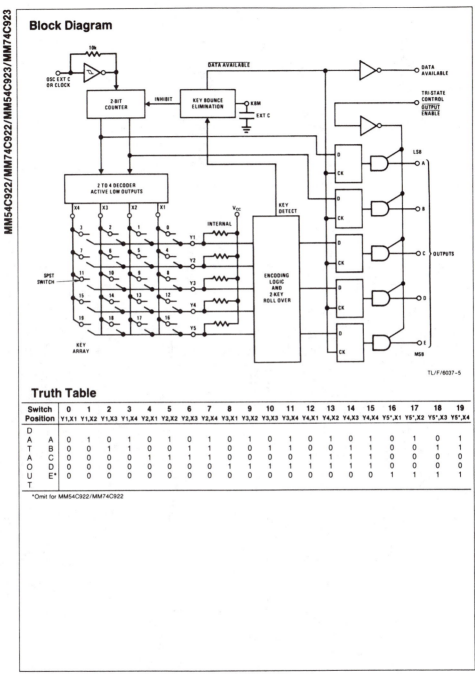

TL/F/6037-5

## Truth Table

| Switch Position | 0 Y1,X1 | 1 Y1,X2 | 2 Y1,X3 | 3 Y1,X4 | 4 Y2,X1 | 5 Y2,X2 | 6 Y2,X3 | 7 Y2,X4 | 8 Y3,X1 | 9 Y3,X2 | 10 Y3,X3 | 11 Y3,X4 | 12 Y4,X1 | 13 Y4,X2 | 14 Y4,X3 | 15 Y4,X4 | 16 Y5*,X1 | 17 Y5*,X2 | 18 Y5*,X3 | 19 Y5*,X4 |
|---|---|---|---|---|---|---|---|---|---|---|---|---|---|---|---|---|---|---|---|---|
| DATA OUT — A | 0 | 1 | 0 | 1 | 0 | 1 | 0 | 1 | 0 | 1 | 0 | 1 | 0 | 1 | 0 | 1 | 0 | 1 | 0 | 1 |
| DATA OUT — B | 0 | 0 | 1 | 1 | 0 | 0 | 1 | 1 | 0 | 0 | 1 | 1 | 0 | 0 | 1 | 1 | 0 | 0 | 1 | 1 |
| DATA OUT — C | 0 | 0 | 0 | 0 | 1 | 1 | 1 | 1 | 0 | 0 | 0 | 0 | 1 | 1 | 1 | 1 | 0 | 0 | 0 | 0 |
| DATA OUT — D | 0 | 0 | 0 | 0 | 0 | 0 | 0 | 0 | 1 | 1 | 1 | 1 | 1 | 1 | 1 | 1 | 0 | 0 | 0 | 0 |
| DATA OUT — E* | 0 | 0 | 0 | 0 | 0 | 0 | 0 | 0 | 0 | 0 | 0 | 0 | 0 | 0 | 0 | 0 | 1 | 1 | 1 | 1 |

*Omit for MM54C922/MM74C922

6-154

**Figure 3.9** Integrated circuit 20-key keyboard scanner MM74C923. (*Source:* Reprinted with permission of National Semiconductor Corporation.)

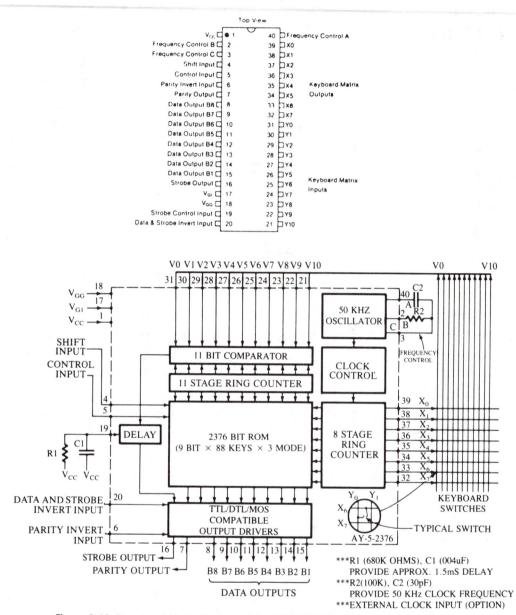

**Figure 3.10** Pinout and block diagram of the AY-5-2376 keyboard encoder. (*Source:* Courtesy of General Instrument Corporation and Microchip Technology Incorporated.)

Hysteresis in a switch means that the voltage needed to turn it on initially is higher than the voltage needed to keep it on. Likewise, the voltage at which the switch turns off is lower than the initial turn-on voltage. An IC device known as a Schmitt trigger gate is commonly used to introduce hysteresis, as shown in Fig. 3.12. The loop inside the gate symbol indicates hysteresis.

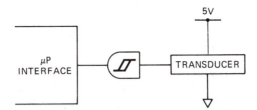

**Figure 3.11** Transistor switch.

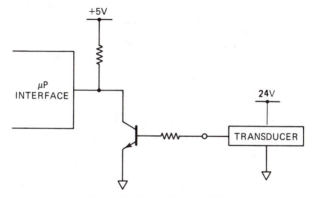

**Figure 3.12** Switching problem solved by hysteresis.

## 3.3.2 Isolation

Isolation is required as a result of this variation of Murphy's law: *If a direct path exists between the outside world and the microprocessor, a destructively high voltage eventually will be applied to it.* The solution is to have a gap in the path through which signals only can be transmitted.

### EXAMPLE 3.6 AN OPTOISOLATOR

An optoisolator (Fig. 3.13) uses an LED and a phototransistor to transmit signals by light while allowing for separate grounds between voltages in the outside world and those inside the microprocessor-based equipment. Optoisolator IC chips can withstand hundreds (even thousands) of volts of difference between input and output ground levels (common mode voltage) and are available with built-in Schmitt triggering.

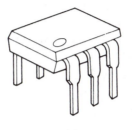

(a)

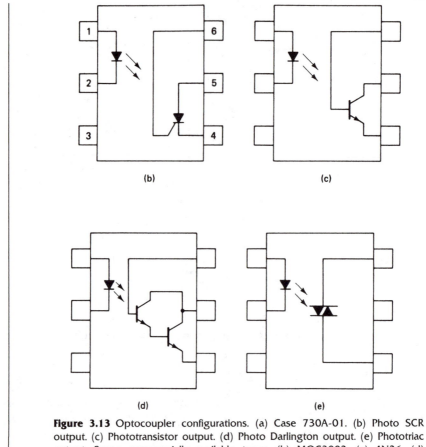

(b)          (c)

(d)          (e)

**Figure 3.13** Optocoupler configurations. (a) Case 730A-01. (b) Photo SCR output. (c) Phototransistor output. (d) Photo Darlington output. (e) Phototriac output. Some commercially available types: (b) MOC3002. (c) 4N26. (d) MOC119. (e) MOC633A. (*Source:* Gayakwad/Sokoloff, *Analog and Digital Control Systems,* © 1988, p. 48. Reprinted by permission of Prentice-Hall, Inc., Englewood Cliffs, N.J.)

Note that isolation is as important for outputs as it is for inputs.

### 3.3.3 Shaft Encoders

A *shaft encoder* is a device that converts the position or rotation of a shaft into a digital signal. There are two basic types: *incremental* and *absolute*. Both types use a disk attached to the shaft. On the disk are concentric strips, and each strip is divided in a binary pattern that can be read by a pickup. The position resolution is determined by the number of divisions (pulses) per rotation.

## EXAMPLE 3.7 INCREMENTAL SHAFT ENCODER

As shown in Fig. 3.14, a typical incremental encoder has three strips (tracks). One track, the index, produces one pulse per revolution. The other two tracks (A and B) produce many pulses per revolution. The two pulse trains are in quadrature, meaning they are 90 degrees apart. Counting the index pulses per minute (tachometry) gives rpm. Counting pulses on the other tracks with respect to the index pulse gives the position of the shaft. Sensing the time sequence of the pulses on the two tracks (A-before-B or B-before-A) gives the direction of rotation. Because of their long life, incremental shaft encoders are often used instead of rotary switches in microprocessor-based equipment.

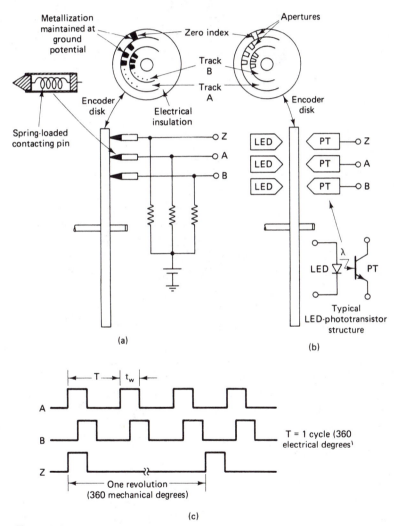

**Figure 3.14** Incremental shaft position encoders. (a) Contacting type. (b) Optical type. (c) Encoder output waveforms. (*Source:* Gayakwad/Sokoloff, *Analog and Digital Control Systems,* © 1988, p. 84. Reprinted by permission of Prentice-Hall, Inc., Englewood Cliffs, N.J.)

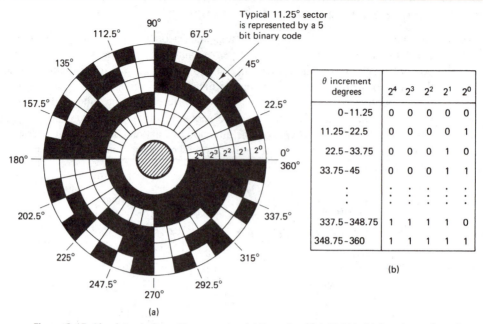

Figure 3.15 shows the table:

| θ increment degrees | $2^4$ | $2^3$ | $2^2$ | $2^1$ | $2^0$ |
|---|---|---|---|---|---|
| 0–11.25 | 0 | 0 | 0 | 0 | 0 |
| 11.25–22.5 | 0 | 0 | 0 | 0 | 1 |
| 22.5–33.75 | 0 | 0 | 0 | 1 | 0 |
| 33.75–45 | 0 | 0 | 0 | 1 | 1 |
| ⋮ | ⋮ | ⋮ | ⋮ | ⋮ | ⋮ |
| 337.5–348.75 | 1 | 1 | 1 | 1 | 0 |
| 348.75–360 | 1 | 1 | 1 | 1 | 1 |

(b)

**Figure 3.15** Absolute shaft position encoder. (a) Encoding disk (5-bit). (b) Angular code assignment truth table. (*Source:* Gayakwad/Sokoloff, *Analog and Digital Control Systems,* © 1988, p. 86. Reprinted by permission of Prentice-Hall, Inc., Englewood Cliffs, N.J.)

Figure 3.15 shows an absolute encoder. Note that for any angular position, the pickups would provide a unique binary number. The more tracks the better the resolution. Absolute encoders are typically more expensive than incremental encoders but have a built-in "memory" of their position. Absolute encoders often use a Grey code instead of simple binary encoding, so that between any two adjacent positions only one bit can change, thus reducing possible errors in reading.

## 3.4 SOLENOIDS AND RELAYS

A *solenoid* is a coil of wire used to produce a magnetic field to move a steel actuator of some sort. When the actuator is used to close a pair of contacts, the device is called a *relay* (Fig. 3.16). Relays and solenoids are common in many kinds of equipment. An impact printer, for example, uses solenoids to drive the printhead and advance the paper. Relays are often used to turn large loads, such as ac motors, on and off.

### 3.4.1 Solenoid Drivers

Two things to know about solenoids: They can be electrically "noisy" and they often require a lot of current. The current requirements mean that most relays and solenoids require a transistor driver. The noise occurs as a short-duration voltage transient, or *spike,* when the current in the coil is abruptly turned off. The spike is due to the collapsing magnetic field inducing a voltage back into the coil, and its amplitude can be many times

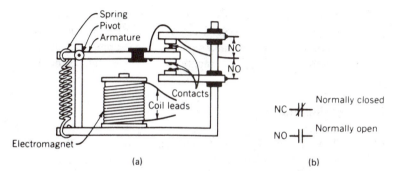

**Figure 3.16** Electromagnetic Relay. (a) Simplified sketch. (b) Symbols showing normal relay contact state. (*Source: Industrial Electronics and Controls* by Martin Newman. Copyright © 1986 John Wiley & Sons. Reprinted by permission of John Wiley & Sons, Inc.)

greater than the supply voltage. Protective circuitry is required to prevent spikes from damaging ("zapping") the transistor driver.

### EXAMPLE 3.8 A DC SOLENOID DRIVER

Figure 3.17 shows a typical solenoid driver for a dc coil, such as a control relay. The diode placed across the coil is the transient protection, as the spike voltage polarity will be opposite the applied voltage. The transistors are shown in a Darlington pair, which has a total current gain that is the product of the individual transistor gains. The gain of a Darlington is typically 2000 or more, so a 1-amp coil could be driven by 0.5 mA of current into the base. Darlingtons are available as integrated circuits.

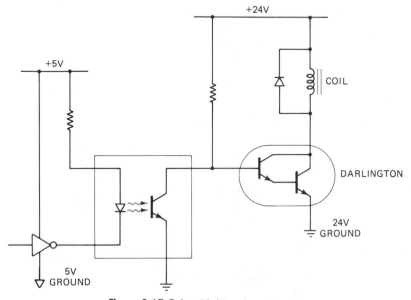

**Figure 3.17** Solenoid driver for a dc coil.

Note that Fig. 3.17 also shows optical isolation. Not only does that provide protection from accidental voltages coming in "backwards" from the output, it also allows the coil to be driven from a separate power supply. Having the microprocessor and digital circuits share a power supply with relays and solenoids is asking for noise problems.

## 3.4.2. AC Solenoids

Many industrial control solenoids and relays operate off the 120V ac mains, usually without transformer isolation. In such cases optical isolation is a necessity. Also, the driver device will be a triac instead of a transistor because a triac can conduct both ways, as required for AC current.

### EXAMPLE 3.9 AN AC SOLENOID DRIVER

Figure 3.18 shows an ac solenoid interface. The resistor-capacitor circuit across the coil is a snubber. It does the same job as the diode across a dc coil, and is often required for larger ac solenoids. The optoisolator differs from what we saw earlier in that it uses a photosensitive triac in place of the phototransistor because the gate of the main triac requires ac drive.

### 3.4.3 Relay Terminology

All that has been said about switch contacts, including bounce, also applies to relays. In addition, relay contacts are sometimes described in terms of their *form:* A N.O. contact pair is called form A; a N.C. pair is called form B; and a SPDT contact arrangement is called form C. The number of contact pairs is specified by a digit, so 1A2B on a relay means it contains a single N.O. pair and two N.C. pairs of contacts. Figure 3.19 shows

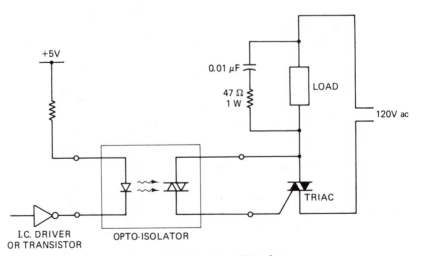

**Figure 3.18** AC solenoid interface.

Chap. 3: Interfacing: Hardware and Software

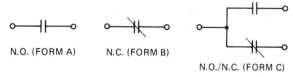

**Figure 3.19** Schematic symbols for relay contacts.

the schematic symbols for relay contacts drawn in their deactivated (i.e., coil de-energized) state.

Relays are divided into ac or dc according to their coil voltage and then further specified by voltage level. Voltages of 5V dc, 12V dc, 24V ac, and 120V ac are common. Note that not all coils of the same voltage draw the same current. Larger relays usually have lower resistance coils.

Relays have some built-in hysteresis. When the coil is energized, the steel pole moves to close the contacts but also shortens the magnetic path. So the pull-in voltage is higher than the drop-out voltage. Additional hysteresis may be needed to prevent *relay chatter,* which is a rapid sequence of relay actuations.

### 3.4.4 Solid State Relays

Solid state relays often combine the isolation, driving, and contact closure functions into a single package. Those meant to control ac loads typically use a triac in place of mechanical contacts. Hybrid designs combine solid state drive with metallic contacts. Solid state relays are fast, quiet, and last indefinitely (if not abused).

### 3.4.5 Software Toggles and Selectors

Many applications use software to make a momentary push-button switch act like a *toggle*. One push turns something on; the next push shuts it off. The program in Example 3.10 implements such a software toggle.

## EXAMPLE 3.10 A SOFTWARE TOGGLE SWITCH

The program here will drive one bit of an output port alternately high and low each time a switch press is detected. The port can be assumed to drive a relay that controls, for example, a motor. DBNCE is a debounce routine, similar to the program in Example 3.3.

```
                ON      EQU     81H             ; DEFINE ON
                OFF     EQU     40H             ; DEFINE OFF
                MOTOR   DATA    P0              ; MOTOR ON PORT O
                SWITCH  DATA    P1              ; SWITCHES ON PORT 1
                FLAG    BIT     78H             ; BIT O OF RAM BYTE 2FH

758040  START:  MOV     MOTOR,#OFF              ; TURN OFF MOTOR
C278            CLR     FLAG                    ; INITIALLY CLEAR FLAG
7590FF          MOV     SWITCH,#0FFH            ; MAKE PINS HIGH, INPUTS
                                                ; ARE CLOSURES TO GROUND

2090FD  LOOP:   JB      SWITCH.0,LOOP           ; LOOP UNTIL SWITCH CLOSES
                                                ; ON BIT O OF PORT 1

111E            ACALL   DBNCE                   ; DEBOUNCE
207807          JB      FLAG,RESET              ; JUMP IF MOTOR ON
D278            SETB    FLAG                    ; ELSE SET FLAG BIT AND
758081          MOV     MOTOR,#ON               ; TURN MOTOR ON
80F1            JMP     LOOP                    ; GO GET NEXT PRESS

C278    RESET:  CLR     FLAG                    ; CLEAR FLAG AND
758040          MOV     MOTOR,#OFF              ; TURN MOTOR OFF
80EA            JMP     LOOP                    ; GO GET NEXT PRESS
```

Another common technique is to simulate a selector switch. Each press of a push button will energize a different "contact" in a fixed sequence.

## EXAMPLE 3.11 A SOFTWARE SELECTOR SWITCH

This program has each press of a switch drive the bits of a port sequentially.

```
                SELECT  DATA    P0              ; OUTPUT  TO PORT O
                SWITCH  DATA    P1              ; INPUT FROM PORT 1
                CLOSED  EQU     0FEH            ; DEFINE CLOSED
                INIT    EQU     01              ; START FROM LSB

75F001          MOV     B,#INIT                 ; INITIALIZE REG B
85F080          MOV     SELECT,B                ; INITIALIZE OUTPUT
7590FF          MOV     SWITCH,#0FFH            ; MAKE PINS HIGH, INPUTS
                                                ; ARE CLOSURES TO GROUND

E590    LOOP:   MOV     A,SWITCH                ; WAIT FOR
B4FEFB          CJNE    A,#CLOSED,LOOP          ; SWITCH PRESS

111A            ACALL   DBNCE                   ; DEBOUNCE IT

E5F0            MOV     A,B                     ; GET SELECTOR
23              RL      A                       ; ROTATE IT LEFT
F5F0            MOV     B,A                     ; SAVE SELECTOR
85F080          MOV     SELECT,B                ; OUTPUT IT
80EF            JMP     LOOP                    ; GO WAIT FOR NEXT
```

## 3.5 DISPLAYS AND PRINTERS

In this section we look at some common display and printing hardware and associated software.

### 3.5.1 LEDs

Light-emitting diodes (LEDs) have all but replaced incandescent lamps on display panels because of their power efficiency and long life. Because they can be turned on and off rapidly, high-power LEDs are sometimes used to drive digital data as pulses of light through fiber-optic cables.

**EXAMPLE 3.12 LED DRIVE**

Figure 3.20a shows a TTL circuit driving a low-power LED. If we assume a 1.6V drop across the LED, the 330-ohm resistor will allow approximately 10 mA of current, a typical level for good brightness. Higher power LEDs will require a transistor (or IC) driver, as shown in Fig. 3.20b.

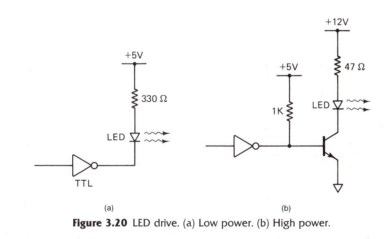

**Figure 3.20** LED drive. (a) Low power. (b) High power.

### 3.5.2 Segment Displays

Seven-segment displays have been implemented in several technologies, including liquid crystal (LCD), fluorescent, and plasma, as well as LED. LED seven-segment displays come in two configurations: *common anode* and *common cathode*, as shown in Fig. 3.21. Each LED corresponds to a segment.

With only seven segments, a display is limited to showing only numeric digits and a few letters, enough for a hexadecimal number. To show a complete alphanumeric character set, a display needs more segments. Figure 3.22 shows a 16-segment display and its character set.

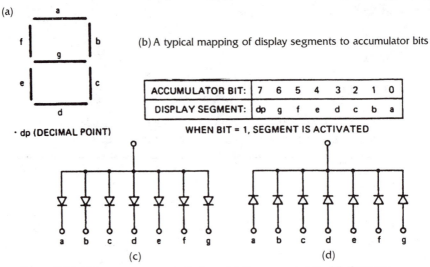

(a)

(b) A typical mapping of display segments to accumulator bits

| ACCUMULATOR BIT: | 7 | 6 | 5 | 4 | 3 | 2 | 1 | 0 |
|---|---|---|---|---|---|---|---|---|
| DISPLAY SEGMENT: | dp | g | f | e | d | c | b | a |

WHEN BIT = 1, SEGMENT IS ACTIVATED

· dp (DECIMAL POINT)

(c)

(d)

**Figure 3.21** LED seven-segment displays. (c) Common anode. (d) Common cathode.

**Figure 3.22** Sixteen-segment display and its character set. (*Source:* Patrick O'Conner, *Digital and Microprocessor Technology,* © 1983, p. 407. Reprinted by permission of Prentice-Hall, Inc., Englewood Cliffs, N.J.)

Segment displays can be driven by connecting each segment to a port bit (through transistor drivers), or they can be driven by *decoder/driver* ICs designed for the purpose. A decoder/driver chip will accept a parallel input (binary or ASCII) and drive the display to show the corresponding character.

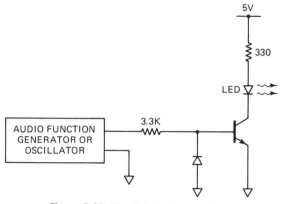

**Figure 3.23** Circuit to test perception.

### 3.5.3 Multiplexed Displays

Suppose a battery-powered instrument has a display of ten seven-segment characters. If each LED uses 10 mA and each segment is lit (all 8s), the total current demand on the battery would be 700 mA just to light the display. In order to conserve power most displays are *multiplexed*.

To multiplex such a display, the first character would be turned on for a brief period of time and then turned off. Then the second character would be flashed on and off; then the third. We would continue until the tenth character was shown and then start the cycle over. Only one seven-segment display would be lit at a time, so the maximum current would be only 70 mA. However, if the multiplexing is done fast enough, human vision integrates the light (smooths it out), so no flickering is seen. Television and motion pictures rely on the same effect. Most people perceive 20 flashes per second or faster as a steady light. To measure your own threshold, build the circuit in Fig. 3.23 and slowly increase the frequency until the flashing appears to give way to a constant light.

Figure 3.24 shows a block diagram of a multiplexed display. Note that a resistor is used for each segment instead of a single resistor for each display. The voltage drop across

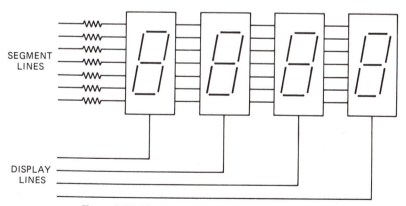

**Figure 3.24** Block diagram of a multiplexed display.

each LED in a display is slightly different, and the lowest one would hog all the current from a single resistor.

## EXAMPLE 3.13 MULTIPLEXED DISPLAY PROGRAM

This program drives a four-digit multiplexed display. The port named DISP is assigned to the display drivers, 1 bit for each digit. Because only 4 digits are used, only the lower 4 bits of the port need be connected. The number held in register B will be used to drive DISP. It is initialized to 11 hex, or 00010001 binary, so every four rotations will cause the sequence to repeat. The numbers to be displayed are held in four consecutive memory locations starting at address ADDR. The subroutine CONVRT converts the binary number to the proper seven-segment equivalent using a look-up table. The port named SEGS drives the segments. Note that all the displays are off when the segments are changed to prevent cross-talk between the displays.

```
              OFF     EQU     0           ; DEFINE OFF
              INIT    EQU     11H         ; SEQUENCE
              DIGITS  EQU     4           ; NUMBER OF DIGITS
              DISP    DATA    P0          ; TO SELECT DISPLAY
              SEGS    DATA    P1          ; TO TURN ON SEGMENTS
              ADDR    DATA    20H         ; DATA BUFFER

   8511F0             MOV     B,INIT      ; SET UP SEQUENCE

   7A04     LOOP:     MOV     R2,#DIGITS  ; NUMBER OF DIGITS
   7820               MOV     R0,#ADDR    ; ADDRESS OF DIGITS

   758000   NEXT:     MOV     DISP,#OFF   ; TURN OFF DISPLAYS
   E6                 MOV     A,@R0       ; GET A DIGIT
   120020             LCALL   CONVRT      ; BINARY TO 7-SEGMENT
   F590               MOV     SEGS,A      ; OUTPUT 7-SEG DIGIT
   08                 INC     R0          ; ADVANCE DIGIT POINTER
   85F080             MOV     DISP,B      ; TURN ON SELECTED DISPLAY
   E5F0               MOV     A,B         ; ADVANCE THE
   23                 RL      A           ;  SEQUENCE TO
   F5F0               MOV     B,A         ;   THE NEXT DISPLAY
   120021             LCALL   DELAY       ; DISPLAY ON-TIME
   DAE9               DJNZ    R2,NEXT     ; GET NEXT DIGIT

   80E3               JMP     LOOP        ; REPEAT 4 DIGIT CYCLE
```

### 3.5.4 Dot-Matrix Characters

As the name implies, a *dot-matrix* character is formed from a two-dimensional array of dots, as shown in Fig. 3.25a. Array sizes are given as width by height, such as $5 \times 7$ or $7 \times 9$. A higher number of dots can approximate fully formed or *near letter quality* (NLQ) characters. In addition to displays, dot-matrix technology is used in printers and other such hard-copy output devices. The dot patterns for the characters are typically stored in a ROM table. Changing character sets (fonts) can be as easy as changing the ROM chip.

## EXAMPLE 3.14  5 × 7 DOT-MATRIX CHARACTERS

As shown in Fig. 3.25, 5 × 7 characters can be stored as 7-bit patterns in 5 consecutive bytes of storage.

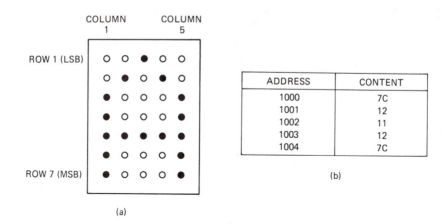

| ADDRESS | CONTENT |
|---------|---------|
| 1000 | 7C |
| 1001 | 12 |
| 1002 | 11 |
| 1003 | 12 |
| 1004 | 7C |

(b)

(a)

**Figure 3.25** Dot-matrix printing. (a) 5 × 7 matrix displaying the letter A. (b) ROM table for storage of dot patterns.

Figure 3.26 shows a generalized dot-matrix printhead. It is a single vertical row because the motion of the head as it is pulled across the paper provides the horizontal dimension. An *impact printer* uses solenoids to press an inked ribbon onto the paper. A *drop-on-demand* type *ink-jet* printer uses piezoelectric crystals (or sometimes solenoids) to shoot drops of ink onto the paper. A *thermal* printer uses heat to ''burn'' each dot onto special paper. An *electrostatic* printer passes pulses of current through special aluminum-coated paper to burn dots.

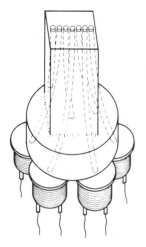

**Figure 3.26** Dot-matrix printhead. (*Source:* Adapted from Patrick O'Connor, *Digital and Microprocessor Technology,* © 1983, p. 424. Reprinted by permission of Prentice-Hall, Inc., Englewood Cliffs, N.J.)

## EXAMPLE 3.15 A DOT-MATRIX DRIVER

The subroutine PRLINE in this program will drive a $5 \times 7$ printhead to print a single line of text terminated by a carriage return (CR). The line is held in a RAM buffer starting at address LBUFF. PRLINE calls CRLF (carriage-return line-feed) to get the printhead to a fresh line. Subroutine LOOKUP will take the ASCII character in the accumulator and return in DPTR the ROM address of the start of the dot character. Subroutine PULSE will print one column of dots, and STEP moves the printhead over to the next column.

```
                HEAD     DATA     P0          ; PRINT HEAD ON PORT 0
                LBUFF    DATA     30H         ; ADDRESS OF LINE BUFFER
                CR       EQU      0DH         ; CARRIAGE RETURN = END OF LINE


  C0E0          PRLINE:  PUSH     ACC         ; SAVE ACCUMULATOR
  C0D0                   PUSH     PSW         ; AND FLAGS

  112F                   ACALL    CRLF        ; NEW LINE
  7830                   MOV      R0,#LBUFF   ; ADDRESS OF LINE BUFFER

  E6            LOOP1:   MOV      A,@R0       ; GET ASCII CHARACTER
  B40D02                 CJNE     A,#CR,CONT  ; END OF LINE? JUMP IF NO
  801C                   SJMP     DONE        ; IF YES, GO HOME

  120032        CONT:    LCALL    LOOKUP      ; GET DOT CHARACTER
  7A05                   MOV      R2,#5       ; SET UP COLUMN COUNT
  75F000                 MOV      B,#0        ; POINTER OFFSET = 0

  E5F0          LOOP2:   MOV      A,B         ; SET UP FOR MOVC
  93                     MOVC     A,@A+DPTR   ; GET A DOT COLUMN
  F580                   MOV      HEAD,A      ; OUTPUT TO PRINT HEAD
  1130                   ACALL    PULSE       ; PRINT A COLUMN
  1131                   ACALL    STEP        ; MOVE HEAD POSITION
  05F0                   INC      B           ; INC POINTER TO NEXT COL
  DAF3                   DJNZ     R2,LOOP2    ; FINISHED THIS CHARACTER?
                                              ;   IF NO, JUMP TO LOOP2
  1131                   ACALL    STEP        ; IF YES, MAKE A SPACE
  1131                   ACALL    STEP        ;   BETWEEN CHARACTERS

  08                     INC      R0          ; NEXT ADDRESS IN BUFFER
  80DE                   JMP      LOOP1       ; GO GET NEXT CHAR FROM BUFFER

  D0D0          DONE:    POP      PSW         ; RESTORE
  D0E0                   POP      ACC         ;   STUFF
  22                     RET                  ; I'M OUT OF HERE
```

## 3.6 HANDSHAKING

*Handshaking* is the exchange of status bits between the processor and an I/O device. Status bits tell such things as whether the device is ready to receive data or if it has data it wishes to send. Often masking will be used to examine a specific bit in a status word and looping is used to wait for the bit to change. A printer, for example, typically has a buffer memory. When the buffer is full, it cannot accept any more characters until some are removed by printing.

Note that the essence of handshaking is an *exchange* of status information. A subroutine may ask a device, "are you ready?" When the device finally says "Yes," the subroutine responds with the (possibly formatted) data and says, "Here it is." The device

Chap. 3: Interfacing: Hardware and Software

takes it, and may then reply, "Thanks, I've got it" to complete the handshake. The signal to "do something" (e.g., print) is called a *strobe*. The signal "OK, it's done" is called an *acknowledgment* (ACK).

## 3.7 CENTRONICS PARALLEL INTERFACE

The parallel interface first used by Centronics Company on its line of inexpensive printers fast became a de facto industry standard.

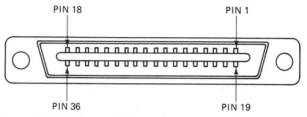

**Figure 3.27** Thirty-six-pin Centronics printer interface connector.

**Table 3.1** Centronics parallel interface signal description.

| Pin | Name | Direction* | Description |
|---|---|---|---|
| 1 | STROBE | to printer | indicates valid data on D1–D8 |
| 2 | D1 | to printer | data line 1 (least significant) |
| 3 | D2 | " | data line 2 |
| 4 | D3 | " | data line 3 |
| 5 | D4 | " | data line 4 |
| 6 | D5 | " | data line 5 |
| 7 | D6 | " | data line 6 |
| 8 | D7 | " | data line 7 |
| 9 | D8 | " | data line 8 (most significant) |
| 10 | ACK | from printer | indicates data was accepted |
| 11 | BUSY | from printer | indicates printer is printing |
| 12 | PE | from printer | indicates "no paper" or "no ribbon" |
| 13 | SLCT | from printer | indicates printer is on-line |
| 14 | GND | none | signal ground |
| 15 | | | no connection |
| 16 | GND | none | signal ground |
| 17 | CHS | none | chassis ground (green wire ground) |
| 18 | VCC | from printer | positive 5 Volts, 20 mA maximum |
| 19–30 | GND | none | signal grounds |
| 31 | RESET | to printer | puts printer into initial state |
| 32 | ERROR | from printer | indicates printing failure or jam |
| 33–36 | | | no connection |

*Note: the above interface is sometimes used bidirectionally

### 3.7.1 Brief Description

As shown in Fig. 3.27, the parallel interface uses a 36-pin connector (e.g., Amphenol part 57-30360) on a ribbon cable up to 15 feet long. It operates at TTL voltage levels. Table 3.1 gives the pin descriptions. Note that the parallel interface can be used for applications other than printers.

### 3.7.2 A Programming Example

The subroutine in Example 3.16 uses the Centronics interface to send a character to a printer each time it is called.

#### EXAMPLE 3.16 CENTRONICS INTERFACE PROGRAM

In the subroutine, port DATA is connected to the data pins (2–9). Port STATUS is connected to $\overline{\text{ERROR}}$, PE, SLCT, BUSY, and $\overline{\text{ACK}}$, with $\overline{\text{ERROR}}$ as the LSB (the upper 3 bits of STATUS are not used). If the status is not "normal," the subroutine returns with CY set. Bit 0 of port STRB is connected to $\overline{\text{STROBE}}$. The strobe pulse must be a minimum of 0.5 µsec long, and the data must remain stable for at least 0.5 µsec before and after the strobe.

```
              FLGBIT  BIT    F0              ; F0 = PSW.5
              DATAP   DATA   P0              ; DATA PORT
              STATUS  DATA   P1              ; STATUS PORT
              STROBE  DATA   P2              ; STROBE ON BIT 0
              MASK    EQU    07              ; FOR LOWER 3 BITS
              NORMAL  EQU    05              ; ERR\ = 1, PE = 0, SLCT = 1

C0E0  PRINT:  PUSH    ACC                    ; SAVE ACCUMULATOR
C0F0          PUSH    B                      ; SAVE REGISTER

7590FF        MOV     STATUS,#0FFH           ; PULL UP THE INPUTS
F5F0          MOV     B,A                    ; SAVE THE CHARACTER
E590          MOV     A,STATUS               ; READ STATUS
5407          ANL     A,#MASK                ; JUST THE LOWER 3 BITS
B40502        CJNE    A,#NORMAL,BAD          ; OK TO PRINT? JUMP IF NO
8007          SJMP    OK                     ; YES, SO HOP TO IT

D2D5  BAD:    SETB    FLGBIT                 ; 1 MEANS CAN'T PRINT

D0F0  BYE:    POP     B                      ; RESTORE REGISTER
D0E0          POP     ACC                    ; RESTORE ACCUMULATOR
22            RET                            ; OUT OF HERE

85F080 OK:    MOV     DATAP,B                ; SEND CHAR TO PRINTER
00            NOP                            ; ALLOW SET-UP TIME
00            NOP                            ; T > 0.5 USEC
C2A0          CLR     STROBE.0               ; FORCE FALLING EDGE
00            NOP                            ; ALLOW TIME
00            NOP                            ; STROBE PULSE LOW
D2A0          SETB    STROBE.0               ; FORCE RISING EDGE

2094FD ACKLP: JB      STATUS.4,ACKLP ; WAIT FOR ACK\
C2D5          CLR     FLGBIT                 ; 0 = PRINTED OK
80E9          JMP     BYE                    ; GO HOME
```

As written, PRINT will not return until the character is acknowledged. It keeps jumping back in a loop until $\overline{\text{ACK}}$ goes low. We should note that, in real life, the printer might fail and the program could wait forever. A better design would wait a finite amount of time and then handle the situation, perhaps with an error message sent to a CRT terminal.

## 3.8 IEEE-488 BUS (GPIB)

Hewlett-Packard Corporation (HP) developed a parallel bus to interconnect its line of programmable instruments. The bus, originally called the Hewlett-Packard Interface Bus (HPIB), became a de facto standard among instrument makers and became known as the General Purpose Instrumentation Bus (GPIB). In 1975, the Institute of Electrical and Electronic Engineers (IEEE) formalized GPIB into the published standard IEEE-488.

### 3.8.1 Brief Description

As shown in Fig. 3.28, the bus consists of 8 bidirectional data lines, 5 bus management lines, 3 handshake lines, and 8 grounds for a total of 24. As many as 15 devices may be connected to the bus. Devices can be separated by no more than 2 m, with a maximum cable length of 20 m. The maximum data transfer rate is 1 million bytes per second. Connections to the bus are by open collector or tristate logic. Because devices are connected to the bus in parallel, all devices must put a high on a line before the line actually goes high. Any device that puts a low on a line pulls that line down for all devices.

Four classes of devices can be hung on the bus:

1. A **Listener** is a device that, when addressed, can only receive data and commands over the bus. An example is a printer. Several listeners can be active simultaneously.
2. A **Talker** is a device that, when addressed, can send data and status over the bus. An example is a digital voltmeter. Only one talker at a time can be active.
3. A **Listener/Talker** is a device that can switch between being a listener and a talker.
4. A **Controller** is a device, typically microprocessor based, that coordinates the activities of the listeners and talkers. There can be more than one controller, but only one can be active at a time.

### 3.8.2 The Three-Wire Handshake

The three handshake signals are as follows:

DAV      **Data Available:** It is asserted high by a talker or controller to indicate that the data on the bus are valid.

NRFD      **Not Ready for Data:** A device pulls NRFD low to indicate that it is not ready to receive. A high means ready.

NDAC      **Not Data Accepted:** A device pulls NDAC low while it is reading data off the bus. When finished, it lets NDAC go high.

HEWLETT-PACKARD INTERFACE BUS

HP-IB

**Figure 3.28** GPIB (IEEE-488) bus interface. (*Source*: Courtesy of Hewlett-Packard Co.)

EXAMPLE 3.17 A HANDSHAKE SEQUENCE

Figure 3.29 shows a typical handshake sequence. A talker puts data on the bus and waits for NRFD to go high. For the NRFD line to go high, all devices must be ready. The talker then asserts DAV and waits for NDAC to go high. Again, all devices that are listening must accept the data before the NDAC line will go high. The talker can then drop DAV.

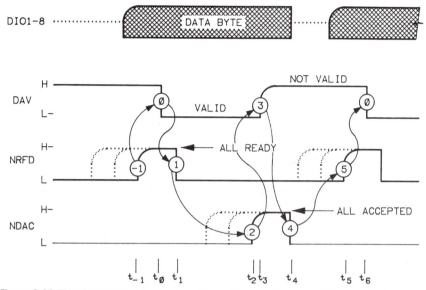

**Figure 3.29** The IEEE-488 three-wire handshake. (*Source:* Courtesy of Hewlett-Packard Co.)

### 3.8.3 Bus Management

Bus management is done with the following signals:

**IFC**   **Interface Clear:** When asserted, it causes devices to cease the current bus activity.

**ATN**   **Attention:** It is asserted when a new talker or controller wishes to take over the bus.

**SRQ**   **Service Request:** A device asserts SRQ to interrupt the current bus activity so it can get service from the controller.

**REN**   **Remote Enable:** It is used to select remote or local control of a device.

**EOI**   **End or Identity:** It is used to identify a device when polled as well as to indicate the end of an operation.

## 3.9 RS-232C (EIA-232) SERIAL INTERFACE

The venerable RS-232 standard (now known as EIA-232) published by the Electronic Industries Association (EIA) is old but still going strong. It is a common data communications physical level (meaning hardware) protocol.

**Table 3.2** RS-232C pin assignment.

| Pin number | Abbreviation | Description |
|:---:|:---:|:---|
| 1 | GWG | Protective chassis ground |
| 2 | TD, SD | Transmit (send) data |
| 3 | RD | Receive data |
| 4 | RTS, RS | Request to send |
| 5 | CTS, CS | Clear to send |
| 6 | DSR | Data set ready |
| 7 | GND | Signal ground |
| 8 | CD, DCD | Carrier detect |
| 20 | DTR | Data terminal ready |
| 22 | RI | Ring indicator |

### 3.9.1 Brief Description

The traditional implementation uses a 25-pin D-type connector (DB25). You will find other types, such as a 9-pin DIN connector, as rarely are all 25 pins used. Table 3.2 gives the descriptions of the commonly used pins. The nominal maximum cable length is 50 feet, although some are longer. Typical voltage levels are ±12V. For data, positive voltage represents binary 0 *(space)* and negative represents binary 1 *(mark)*. For the status pins, positive is true (on) and negative is false (off). When no data are being sent, the data line is held high (kept *marking*).

A complete description of RS-232C is not given here because it can be found in any data communications text. Briefly, data are sent, 1 bit at a time (serially), over an RS-232C link in the form of asynchronous frames. *Asynchronous* simply means that the timing between the frames can vary. As shown in Fig. 3.30, each frame consists of a start bit (always low), data bits, a parity bit, and a stop bit (always high). Data are sent LSB first. The rate at which the frame bits are sent is called the *baud rate* (or just baud) and is essentially the same as bits per second (bps). Note that while one stop bit is typical, one

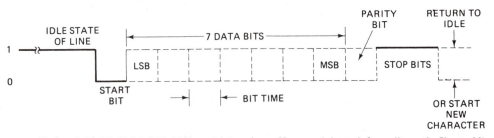

**Figure 3.30** RS-232C (EIA-232) serial interface. (*Source:* Adapted from Kenneth Short, *Microprocessors and Programmed Logic,* © 1981, p. 282. Reprinted by permission of Prentice-Hall, Inc., Englewood Cliffs, N.J.)

and a half or two stop bits are used in certain applications (e.g., low baud electro-mechanical equipment).

Common baud rates are 300, 600, 1200, 2400, 4800, and 9600. A frame may contain 5, 6, 7, or 8 data bits. The parity bit is optional, and may be odd or even. Considering the many variations on RS-232, it may be wise to select a printer with a parallel interface. Modems especially require RS-232.

### 3.9.2 A Programming Example

The following shows how to use the 8051 serial port in a polling mode.

---

**EXAMPLE 3.18 SERIAL DATA**

This subroutine will send a serial frame when called. SOD would be connected to a line driver chip to convert the TTL voltage to RS-232C levels. The basic handshake is for the processor to *assert* (i.e., make true) RTS and wait for CTS as the go-ahead to send.

```
                    ;    ASSUMES THAT XTAL FREQ = 11.981 MHZ
                    ;    BAUD RATE = 1200 BPS
                    ;    SEE EXAMPLE 2.1

            COUNT   EQU     230             ; COUNT FOR 1200 BPS
            EOT     EQU     04H             ; END OF TRANSMIT
            SMOD    BIT     87H.7           ; PCON.7
            RTS     BIT     P0.0            ; ACTIVE LOW TO MODEM
            CTS     BIT     P0.1            ; ACTIVE LOW FROM MODEM
            DSR     BIT     P0.2            ; ACTIVE LOW FROM MODEM

75A800 START:  MOV     IE,#00          ; DISABLE INTERRUPTS
7580FF         MOV     P0,#0FFH        ; PULL-UP PORT 0 PINS
C28E           CLR     SMOD            ; SELECT DIVIDE BY 384
C28C           CLR     TR0             ; TURN OFF TIMER 0
C28E           CLR     TR1             ; TURN OFF TIMER 1

758920         MOV     TMOD,#20H       ; TIMER 1 IN MODE 2
758DE6         MOV     TH1,#COUNT      ; LOAD REGISTERS
758BE6         MOV     TL1,#COUNT      ;   FOR COUNT-DOWN
D28E           SETB    TR1             ; TURN ON TIMER 1
759860         MOV     SCON,#60H       ; SERIAL MODE 1
                                       ; TRANSMIT ONLY, REN = 0
2082FD LP1:    JB      DSR,LP1         ; WAIT FOR DSR
C280           CLR     RTS             ; ASSERT RTS
2081FD LP2:    JB      CTS,LP2         ; WAIT FOR CTS

120049 SLP:    LCALL   GETCHAR         ; RETURNS CHAR IN ACC
B40402         CJNE    A,#EOT,CONT     ; CONTINUE IF NOT EOT
8009           SJMP    RCVE            ; END OF TRANSMIT
F599   CONT:   MOV     SBUF,A          ; SEND CHAR TO UART
3099FD LP3:    JNB     TI,LP3          ; WAIT UNTIL CHAR SENT
C299           CLR     TI              ; CLEAR TRANSMIT FLAG
80EF           JMP     SLP             ; SEND NEXT CHAR

C298   RCVE:   CLR     RI              ; INITIAL CLEAR
D29C           SETB    REN             ; ENABLE RECEIVER

3098FD RLP:    JNB     RI,RLP          ; WAIT FOR CHAR
E599           MOV     A,SBUF          ; GET CHAR FROM UART
B40402         CJNE    A,#EOT,CONR     ; CONTINUE IF NOT EOT
8007           SJMP    BYE             ; OTHERWISE, GO AWAY
12004A CONR:   LCALL   PUTCHAR         ; STORE RECEIVED CHAR
C298           CLR     RI              ; CLEAR RECEIVE FLAG

80EF           JMP     RLP             ; GO GET NEXT CHAR

00     BYE:    NOP                     ; REST OF PROGRAM
```

---

### 3.9.3 Checksums

When data are transmitted any distance, it is inevitable that errors will creep in. If you can't prevent it, you must detect and correct it. A full treatment of error detection and correction is beyond the scope of this text; however, we will look at the simple but very effective technique of *checksums*.

In comparison to simple parity, which is based on a single frame, checksums are based on a *block*. To implement checksums, the data are broken down into fixed-size blocks. For example, a 2K data file could be divided into eight blocks of 256 bytes each. All the bytes in a block are added up, even if they are ASCII characters, to get a 16-bit total. The addition is *modulo,* meaning you don't worry about overflow. The checksum is then formed by taking the two's complement of the total.

The checksum is sent as 2 bytes at the end of each block. As the receiver receives the data, it generates its own version of the checksum and compares it to the 2 bytes it received at the end of the block. If they agree, the block is assumed to be good. Otherwise, the block is bad and must be retransmitted. The receiver will send either a positive or negative acknowledgment (ACK or NAK) back to the transmitter after each block. (Actually, to verify the checksum the receiver just has to add up all the received bytes and then add the checksum. If the total is zero, the checksum is valid and the block is good.)

---

**EXAMPLE 3.19 CHECKSUM CALCULATION**

To simplify the arithmetic, let's find an 8-bit checksum for four 4-bit words.

|  |  | one's complement of total is:   11010100 |
|---|---|---|
| word 1 = | 1001 | |
| word 2 = | 0110 | |
| word 3 = | 1111 | two's complement of total is the check- |
| word 4 = | 1101 | sum:   11010101 |
| total = | 00101011 | |

| Verification: | 1001 | |
|---|---|---|
| | 0110 | |
| | 1111 | received data |
| | 1101 | |
| | 00101011 | sum of received data |
| + | 11010101 | add the checksum |
| | 00000000 | total is zero; data are good |

## 3.10 SOME PARALLEL I/O APPLICATIONS

The following are some typical applications.

### 3.10.1 Stepper Motor Sequencer

Most motors spin continuously when voltage is applied. As the name implies, stepper motors move in discrete steps. A full treatment of stepper motors can be found in Gayakwad and Sokoloff. Briefly, the armature of a stepper is surrounded by a number of coils. As the coils are energized in a certain sequence, the armature will move in steps of a fixed angular size. The faster the sequence is repeated, the faster the motor turns. When the coils are held energized in a fixed state, the motor is held (locked) in position.

### EXAMPLE 3.20 STEPPER MOTOR SEQUENCER PROGRAM

Figure 3.31a shows the schematic of a four-phase (four-coil) motor and the associated coil drive sequence. If the sequence is executed from top to bottom, the motor turns clockwise (CW); bottom to top execution gives CCW rotation. Steppers are specified by steps per revolution (e.g., 200 steps/rev), as well as by speed and holding torque.

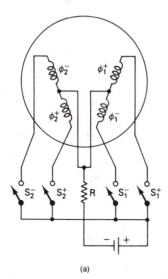

| Step | $S_1^+$ | $S_1^-$ | $S_2^+$ | $S_2^-$ | HEX |
|------|---------|---------|---------|---------|-----|
| 1 | X | | | X | 9 |
| 2 | | X | | X | 5 |
| 3 | | X | X | | 6 |
| 4 | X | | X | | A |
| 1 | X | | | X | 9 |

(a)  (b)

**Figure 3.31** Stepper motor sequencer. (a) Schematic of a four-phase (four-coil) motor. (b) Associated coil drive sequence. (*Source:* Adapted from Gayakwad/Sokoloff, *Analog and Digital Control Systems*, © 1988, p. 153. Reprinted by permission of Prentice-Hall, Inc., Englewood Cliffs, N.J.)

```
; P1.6 - P1.0 =  NUMBER OF STEPS (00H - 7FH)
; P1.7 = DIRECTION:  P1.7 = 0 -> FORWARD,  P1.7 = 1 -> REVERSE

                S_1     EQU     09              ; SEQ = 1001
                S_2     EQU     05              ; SEQ = 0101
                S_3     EQU     06              ; SEQ = 0110
                S_4     EQU     0AH             ; SEQ = 1010
                NSQ     EQU     5               ; NUMBER OF SEQS + 1

C2D4    START:  CLR     RS1             ; SELECT REG BANK 1
D2D3            SETB    RS0             ;   08H - 0FH
853081          MOV     SP,30H          ; INITIALIZE STACK POINTER

7590FF          MOV     P1,#0FFH        ; PULL UP FOR INPUT
758009          MOV     P0,#S_1         ; INITIALIZE MOTOR
75F000          MOV     B,#0            ; INITIAL OFFSET
12004E          LCALL   DELAY           ; WAIT 1 STEP TIME
90004A          MOV     DPTR,#TABLE     ; GET TABLE ADDRESS

E590    MLP:    MOV     A,P1            ; READ STEP COUNT
547F            ANL     A,#7FH          ; JUST LOWER 7 BITS
60FA            JZ      MLP             ; IGNORE ZERO STEPS
FA              MOV     R2,A            ; SAVE THE COUNT
E5F0            MOV     A,B             ; GET OFFSET
8597D5          MOV     F0,P1.7         ; SAVE DIRECTION
20D504          JB      F0,REV          ; FORWARD OR REVERSE?
112D            ACALL   F_MOVE          ; CALL FORWARD DRIVER
80ED            JMP     MLP             ; DO IT AGAIN
1135    REV:    ACALL   R_MOVE          ; CALL REVERSE DRIVER
80E9            JMP     MLP             ; DO IT AGAIN

04      F_MOVE: INC     A               ; FORWARD 1 SEQ
B4050A          CJNE    A,#NSQ,OK       ; PAST THE END?
7400            MOV     A,#0            ; YES: WRAP AROUND
8006            SJMP    OK              ; SKIP OVER R_MOVE
14      R_MOVE: DEC     A               ; BACK 1 SEQ
B4FF02          CJNE    A,#0FFH,OK      ; PAST THE END?
7404            MOV     A,#4            ; YES: WRAP AROUND
F5F0    OK:     MOV     B,A             ; SAVE FOR NEXT TIME
93              MOVC    A,@A+DPTR       ; GET THE SEQUENCE
F580            MOV     P0,A            ; SEND IT TO STEPPER
114E            ACALL   DELAY           ; WAIT 1 STEP TIME
DA01            DJNZ    R2,SLP          ; NEXT STEP OR DONE?
22              RET                     ; DONE: BACK TO MLP
20D5ED  SLP:    JB      F0,R_MOVE       ; LOOP FOR REVERSE
80E3            JMP     F_MOVE          ; LOOP FOR FORWARD

09      TABLE:  DB      S_1             ; SEQ 1
05              DB      S_2             ; SEQ 2
06              DB      S_3             ; SEQ 3
0A              DB      S_4             ; SEQ 4
```

### 3.10.2 Strobed I/O: A Printer Example

In the program in Example 3.21 data will be read from an ASCII keyboard and sent to a printer over a parallel interface, as shown in Fig. 3.32.

**EXAMPLE 3.21 STROBED I/O**

```
          ; EXTERNAL INTERRUPT PIN INTO (P3.2) WILL BE SET UP
          ; FOR FALLING EDGE ACTIVATION BY DAV.  EDGE ON INTO
          ; IS LATCHED INTO BIT IEO (TCON.1) FOR POLLING

                KBD     DATA    P1      ; DATA FROM KEYBOARD
                PRNTR   DATA    PO      ; DATA TO PRINTER
                STROBE  BIT     P2.0    ; PRINTER STROBE
                ACK     BIT     P2.1    ; ACK\ FROM PRINTER
                DAV     BIT     P3.2    ; DAV\ ON INTO

75B000  START:  MOV     P3,#0           ; NO OTHER ALT FUNCS
D288            SETB    ITO             ; SET FOR FALLING EDGE
C289            CLR     IEO             ; INITIAL CLEAR
C2AF            CLR     EA              ; NO INTERRUPTS
D2AO            SETB    STROBE          ; PULL UP
D2A1            SETB    ACK             ;   THESE
D2B2            SETB    DAV             ;    PORT LINES

3089FD  KBLP:   JNB     IEO,KBLP        ; WAIT FOR DAV\
C289            CLR     IEO             ; RESET IEO
859080          MOV     PRNTR,KBD       ; TRANSFER DATA
OO              NOP                     ; SET-UP TIME FOR DATA
C2AO            CLR     STROBE          ; ASSERT STROBE\
OO              NOP                     ; STRETCH IT OUT
D2AO            SETB    STROBE          ; PULL STROBE\ HIGH
20A1FD  ACKLP:  JB      ACK,ACKLP       ; WAIT FOR ACK\
80ED            JMP     KBLP            ; DO IT AGAIN
```

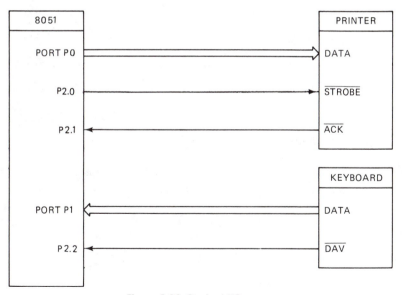

**Figure 3.32** Strobed I/O.

## 3.11 SOME INTERRUPT-DRIVEN APPLICATIONS

In programs like that in Example 3.21, the computer spends all its time in a polling loop waiting for something to happen. Because I/O is often a rare event (in the time frame of a CPU), we could let the processor do something else until an I/O device demands service through an interrupt. However, interrupts must be used with care because they can cause intermittent bugs that are hard to track down.

### 3.11.1 Interrupt I/O: An Example

The programmed I/O of Example 3.21 could be changed into an *interrupt service routine*.

**EXAMPLE 3.22 INTERRUPT I/O**

```
        ; EXTERNAL INTERRUPT PIN INTO (P3.2) WILL BE SET UP
        ; FOR FALLING EDGE ACTIVATION BY DAV.  FALLING EDGE
        ; ON INTO WILL GENERATE AN INTERRUPT

             KBD     DATA    P1       ; DATA FROM KEYBOARD
             PRNTR   DATA    PO       ; DATA TO PRINTER
             STROBE  BIT     P2.0     ; PRINTER STROBE
             ACK     BIT     P2.1     ; ACKNOWLEDGE
             DAV     BIT     P3.2     ; DAV ON INTO

             ORG     03               ; INTO VECTOR ADDRESS
859080       MOV     PRNTR,KBD        ; TRANSFER DATA
00           NOP                      ; SET-UP TIME FOR DATA
C2A0         CLR     STROBE           ; ASSERT STROBE\
00           NOP                      ; STRETCH IT OUT
D2A0         SETB    STROBE           ; PULL STROBE\ HIGH
20A1FD  ACKLP: JB    ACK,ACKLP        ; WAIT FOR ACK\
32           RETI                     ; RETURN

             ORG     40H              ; ABOVE INTERRUPTS
75B000  START: MOV   P3,#0            ; NO OTHER ALT FUNCS
D288         SETB    ITO              ; SET FOR FALLING EDGE
C289         CLR     IEO              ; INITIAL CLEAR
75A800       MOV     IE,#0            ; CLEAR IT ALL
D2AF         SETB    EA               ; MASTER ENABLE
D2A8         SETB    IE.O             ; ENABLE INTO
D2A0         SETB    STROBE           ; INITIALIZE
D2A1         SETB    ACK              ;   THESE BITS
D2B2         SETB    DAV              ;    HIGH
00           NOP                      ; REST OF PROGRAM
                                      ; STARTS HERE
```

### 3.11.2 A Real-Time Clock (RTC)

Basically, an RTC keeps accurate track of time. Peripheral devices such as the National Semiconductor MM58274 (Fig. 3.33) are available that will give hours, minutes, seconds, and tenths of seconds, as well as month and day of the week. Sometimes it is necessary to implement a simple RTC in software. For example, an industrial control program may need to monitor parameters such as temperature and pressure periodically. In such a case the "ticks of the clock" are usually interrupts generated by a *time base*, such as a crystal oscillator or even the 60-Hz (50-Hz in Europe) ac line (suitably isolated, of course).

Sometimes the RTC function is part of another interrupt routine. For example, a system may want to poll all its inputs once every 10 sec. An interrupt every second would

**National Semiconductor Corporation**

May 1987

# MM58274B Microprocessor Compatible Real Time Clock

## General Description

The MM58274B is fabricated using low threshold metal gate CMOS technology and is designed to operate in bus oriented microprocessor systems where a real time clock and calendar function are required. The on-chip 32.768 kHz crystal controlled oscillator will maintain timekeeping down to 2.2V to allow low power standby battery operation. This device is pin compatible with the MM58174A but continues timekeeping up to tens of years. The MM58274B is a direct replacement for the MM58274 offering improved Bus access cycle times.

## Applications

- Point of sale terminals
- Teller terminals
- Word processors
- Data logging
- Industrial process control

## Features

- Same pin-out as MM58174A and MM58274
- Timekeeping from tenths of seconds to tens of years in independently accessible registers
- Hours counter programmable for 12 or 24-hour operation
- Leap year register
- Buffered crystal frequency output in test mode for easy oscillator setting
- Data-changed flag allows simple testing for time rollover
- Independent interrupting time with open drain output
- Fully TTL compatible
- Low power standby operation (10μA at 2.2V)
- Low cost 16-pin DIP and 20-pin PCC

## Block Diagram

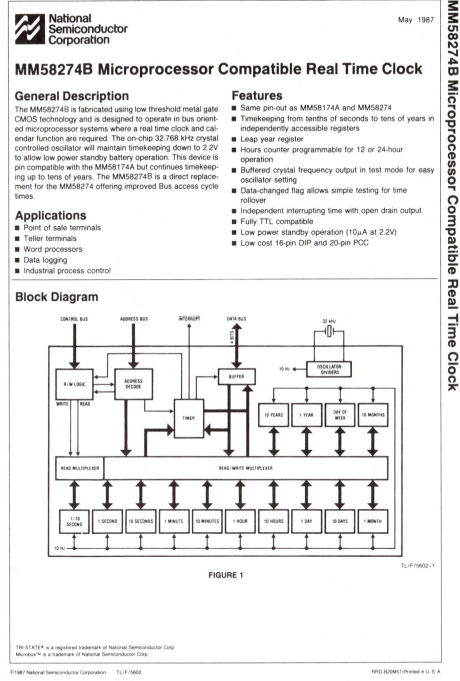

FIGURE 1

TL/F/5602–1

**Figure 3.33** National Semiconductor MM58274.

invoke the service routine, which would increment the clock and do the poll every tenth interrupt. Also, multitasking systems use an interrupt RTC to determine when to activate (wake up) and deactivate programs (tasks).

## EXAMPLE 3.23 REAL-TIME CLOCK

```
                SO1     DATA    30H             ; SECONDS UNITS
                S10     DATA    SO1 + 1         ; SECONDS TENS
                MO1     DATA    SO1 + 2         ; MINUTES UNITS
                M10     DATA    SO1 + 3         ; MINUTES TENS
                START   DATA    40H             ; STACK GROWS UPWARD

        ; ASSUMES A 1 HZ CLOCK IS DRIVING INT1 PIN (P3.3)

                ORG     13H             ; INT1 VECTOR ADDRESS
0140            AJMP    RTC             ; JUMP TO START OF RTC

                ORG     40H             ; ABOVE INTERRUPTS
COEO    RTC:    PUSH    ACC             ; SAVE ACCUMULATOR
0530            INC     SO1             ; INC UNITS OF SECS
E530            MOV     A,SO1           ; GET FOR COMPARE
B40A21          CJNE    A,#10,BYE       ; PAST 9? JUMP IF NO
753000          MOV     SO1,#0          ; IF YES, S_UNITS = 0
0531            INC     S10             ; INC TENS OF SECS
E531            MOV     A,S10           ; GET FOR COMPARE
B40617          CJNE    A,#6,BYE        ; PAST 5? JUMP IF NO
753100          MOV     S10,#0          ; IF YES, S_TENS = 0
0532            INC     MO1             ; INC UNITS OF MINS
E532            MOV     A,MO1           ; GET FOR COMPARE
B40A0D          CJNE    A,#10,BYE       ; PAST 9? JUMP IF NO
753200          MOV     MO1,#0          ; IF YES, M_UNITS = 0
0533            INC     M10             ; INC TENS OF MINS
E533            MOV     A,M10           ; GET FOR COMPARE
B40603          CJNE    A,#6,BYE        ; PAST 5? JUMP IF NO
753300          MOV     M10,#0          ; IF YES, M_TENS = 0
DOEO    BYE:    POP     ACC             ; RESTORE ACCUMULATOR
32              RETI                    ; RETURN

                ORG     100H            ; MAIN STARTS HERE
758140  MAIN:   MOV     SP,#START       ; SET-UP STACK POINTER
75B000          MOV     P3,#0           ; NO OTHER SPECIAL FUNCS
D2B3            SETB    INT1            ; PULL UP P3.3 = INT1\
D28A            SETB    IT1             ; SET FOR FALLING EDGE
C28B            CLR     IE1             ; INITIAL CLEAR
75A800          MOV     IE,#0           ; CLEAR IT ALL
D2AF            SETB    EA              ; MASTER ENABLE
D2AA            SETB    EX1             ; IE.2 = 1 ENABLES INTO
00              NOP                     ; REST OF PROGRAM
                                        ;   STARTS HERE
C2AA            CLR     EX1             ; DISABLE INT1\
E531            MOV     A,S10           ; READ TENS OF SECS
8530F0          MOV     B,SO1           ; READ UNITS OF SECS
D2AA            SETB    EX1             ; RE-ENABLE INT1\
311F            ACALL   TIMER           ; DO SOMETHING WITH TIME
                                        ; REST OF PROGRAM
```

Note that when the main program wants to read the time it must first disable interrupts. Otherwise, an error could occur as follows: Suppose the time is 49 sec and the main program reads the 4. But an interrupt occurs before the 9 is read, so the second digit will be changed to 0 before the main routine gets a chance to read it. After the interrupt, the program reads the next digit and gets a time of 40 sec instead of the correct value.

Chap. 3: Interfacing: Hardware and Software

A section of code that cannot be interrupted is often called a *critical region,* and is an example of the sort of synchronization required when programs are run *concurrently.* The RTC and the program that reads the time are concurrent; in effect, they are running at the same time even though they share a single processor. A problem with using critical regions with an RTC is that, unless an interrupt is latched or held pending while in the region, the clock may miss ticks and slowly fall behind the correct time.

### 3.11.3 Serial I/O with Interrupts

The program of Example 3.24 uses the 8051 in serial mode 1 to send and receive simultaneously asynchronous data at a baud rate of 1200 bits per second (bps). We will use timer 1 in auto-reload mode (timer mode 2) to generate the bit clock. We will use a CPU clock frequency of 11.059 MHz because that is a multiple of the baud rate.

### EXAMPLE 3.24 SERIAL I/O

The serial port generates an interrupt when either a frame is received or when a frame being sent gets to the last (stop) bit. The interrupt service routine must determine which event occurred by testing both the TI and RI flag bits. TI and RI are set by the hardware when the corresponding interrupt occurs but must be reset in software as part of the service routine.

```
            PCON    EQU    87H           ; NOT PREDEFINED IN MY ASSEMBLER
            COUNT   EQU    0E8H          ; TIMER RELOAD VALUE
            SFLG    BIT    20H.O         ; SEND FLAG
            DONE    EQU    01            ; DEFINE DONE

            ORG    0000                  ; POWER UP TO HERE
804E                SJMP   MAIN          ; GO TO MAIN

            ORG    0023H                 ; SERIAL INTERRUPT VECTOR
8000                SJMP   SIR           ; JUMP TO INTERRUPT ROUTINE

309904      SIR:    JNB    TI,NXT        ; JUST SENT FRAME? JUMP IF NO
C299                CLR    TI            ; YES, CLEAR XMIT INTERRUPT FLAG
C200                CLR    SFLG          ; AND TELL THE SUBROUTINE
309806      NXT:    JNB    RI,BYE        ; JUMP IF NO FRAME RECEIVED
E599                MOV    A,SBUF        ; GET THE 8-BIT DATA WORD
F6                  MOV    @R0,A         ; AND SAVE IT IN BUFFER
08                  INC    R0            ; INC POINTER TO BUFFER
C298                CLR    RI            ; CLEAR RCVE INTERRUPT FLAG
32          BYE:    RETI                 ; RETURN FROM INTERRUPT

            ORG    50H                   ; MAIN STARTS HERE
75A800      MAIN:   MOV    IE,#00        ; DISABLE INTERRUPTS
854081              MOV    SP,40H        ; MOVE SP TO HIGH RAM
85E88D              MOV    TH1,COUNT     ; INIT COUNT
85E88B              MOV    TL1,COUNT     ; FOR BAUD RATE
758920              MOV    TMOD,#20H     ; INIT TIMER 1 MODE 2
758840              MOV    TCON,#40H     ; INIT TIMER CONTROL
759870              MOV    SCON,#70H     ; SERIAL MODE 1, 10 BITS/FRAME
758700              MOV    PCON,#00H     ; SET SMOD TO 0
75A890              MOV    IE,#90H       ; ENABLE SERIAL INTERRUPTS
7820                MOV    R0,#BUFR      ; ADDRESS OF BUFR IN R0
E547                MOV    A,'G'         ; GET FIRST ASCII CHAR
120077              LCALL  SEND          ; AND SEND IT
E54F                MOV    A,'O'         ; GET SECOND ASCII CHAR
120077              LCALL  SEND          ; AND SEND IT
```

```
          ;                    <REST OF MAIN PROGRAM>

F599      SEND:   MOV     SBUF,A        ; LOAD OUTPUT BUFFER
D200              SETB    SFLG          ; SET FLAG TO BUSY STATUS
2000FD    SLP:    JB      SFLG,SLP      ; WAIT FOR INTERRUPT TO
22                RET                   ; RETURN FROM SUBROUTINE

                  DSEG                  ; DATA SEGMENT IN RAM
                  ORG     20H           ; BUFFER ADDRESS IN RAM
          BUFR:   DS      16            ; RECEIVED DATA STORAGE
```

Note that the subroutine can be anywhere in the program address space because we called it with the 3-byte LCALL instruction. If the subroutine is within a 2K range, we can use the 2-byte ACALL. Many assemblers have a generic CALL instruction that automatically gets replaced with either ACALL or LCALL, as is appropriate; the assembler computes the address range. Also note that subroutine returns use RET, whereas interrupt returns use RETI.

## 3.12 ANALOG/DIGITAL INTERFACING

Microprocessors process digital numbers, but most of the things we want to measure and control, things out in the real world, are *analog*. Examples are pressure in a boiler, speed of a motor, and position of a robot arm. They are analog because they can take on any value within their range and can change smoothly and continuously from one value to another. Even though we use transducers to convert such things to an electrical signal, those signals vary in the same manner as (i.e., they are analogs of) the actual thing itself. On the other hand, digital numbers can have only discrete values in a range determined by the number of bits in the digital word. For example, an 8-bit binary number can go from 0 to 255 *full scale* and has a *resolution* of 1 LSB or approximately 0.4 percent of full scale (1/255).

What we need are analog to digital converters (A/D or ADC) and digital to analog converters (D/A or DAC). Such devices can be characterized in terms of the following: *speed* (how many times per second can it convert; also called sample rate), *resolution* (how many bits does it have), *accuracy* (how many bits are correct), *linearity* (how close does output track input), and *monotonicity* (does the output always go in the same direction as the input). The digital end of DACs and ADCs is usually pure binary, but may also be BCD or some other code.

ADCs and DACs can be connected to the CPU through I/O ports, or, if they have the appropriate interface, they can be connected to the CPU bus lines and treated as memory locations (i.e., they can be memory mapped). Microprocessor-compatible devices have the bus interface built in.

### 3.12.1 Digital to Analog Converters (DACs)

The output of a DAC can be current or voltage. It can be *unipolar* (e.g., 0 to +10V) or *bipolar* (e.g., −5 to +5V). Output voltages are derived from a *reference,* which can be either built-in or external. If the reference can be varied, it is a *multiplying DAC,* where the output is proportional to the binary input times the reference voltage. The speed of the DAC, often specified as a *settling time,* determines how often the program can send data to it. For example, a 10-μsec settling time means 100,000 conversions/sec (1 conversion/10 μsec).

Interfacing a DAC to a microprocessor is relatively straightforward. Basically, it has a number of input lines for the bits and an analog output line. There may be separate analog and digital ground lines. Refer to the simplified diagram in Fig. 3.34. If it is *microprocessor compatible,* it is simply connected to the bus lines and written to by moves to the appropriate "memory" addresses. If the DAC is connected to an I/O port, there are two things to consider:

**First,** the binary input must be *latched* to hold it constant while the DAC converts it to analog. If the DAC contains its own latch, then the microprocessor can use a nonlatched output port. Data are written to the port and the latch-strobe line on the DAC is pulsed to latch it. The data must not change during the strobe pulse.

**Second,** the DAC word size may exceed the CPU word size. For example, the ports may be 8 bits while the DAC is 12 bits. Because two successive OUT instructions will be needed to send data to the DAC, the DAC output will be at some arbitrary value (a glitch) in between the two OUTs. In such a case *double buffering* can be used. As

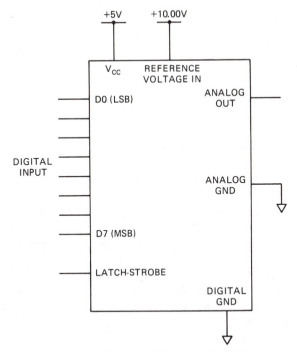

**Figure 3.34** Eight-bit DAC with built-in latch.

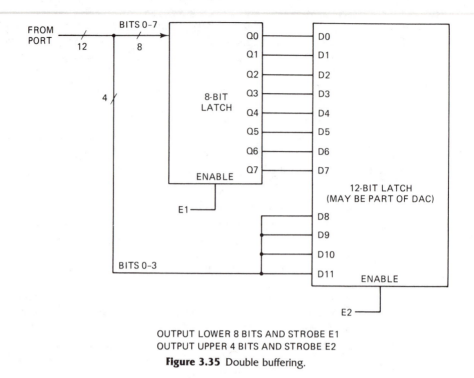

BITS 0-7

FROM PORT /12 /8

4 /

BITS 0-3

8-BIT LATCH

Q0
Q1
Q2
Q3
Q4
Q5
Q6
Q7

ENABLE

E1

D0
D1
D2
D3
D4
D5
D6
D7

12-BIT LATCH
(MAY BE PART OF DAC)

D8
D9
D10
D11

ENABLE

E2

OUTPUT LOWER 8 BITS AND STROBE E1
OUTPUT UPPER 4 BITS AND STROBE E2

**Figure 3.35** Double buffering.

shown in Fig. 3.35, double buffering allows the CPU to output the first part of the 12-bit word to an 8-bit "holding area." It then outputs the second part. After both parts are in place, it strobes the 12-bit latch so that the DAC sees the new word "all in one piece."

### 3.12.2 Pulse Width Modulator (PWM) D/A

Many IC devices, such as the DAC0808 shown in Fig. 3.36, are available for digital to analog conversion. We can also implement a cheap, simple, slow-speed DAC by using a single I/O bit to output a constant frequency pulse train and integrating it with a low-pass filter, as shown in Fig. 3.37. By digitally varying the duty cycle (on/off ratio) of the pulse, the analog dc voltage out of the filter will be proportional to the digital input.

We will use two 8-bit numbers in this routine: the binary number to be converted and the duration count (initialized to 255). We will initialize the I/O bit high and set up the timer to generate an interrupt every 100 μsec (assuming a 12-MHz clock). Remember that in timer mode, the clock is automatically divided by 12 before it is used.

The interrupt service routine will decrement both the binary number and the duration count. When the binary number is decremented to zero, the I/O bit is forced low. The duration count continues to be decremented until it reaches zero, at which point the I/O bit is forced high, the duration count is reset to 255, and the binary number is restored. The result is a pulse train with a constant period of 25.5 msec and a duty cycle that varies with the binary number.

# Block and Connection Diagrams

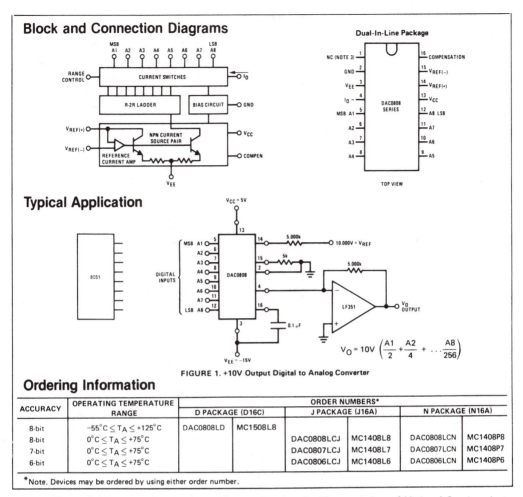

## Typical Application

FIGURE 1. +10V Output Digital to Analog Converter

$$V_O = 10V \left( \frac{A1}{2} + \frac{A2}{4} + \ldots \frac{A8}{256} \right)$$

## Ordering Information

| ACCURACY | OPERATING TEMPERATURE RANGE | ORDER NUMBERS* | | | | |
|----------|-----------------------------|----------------|----------------|----------------|----------------|----------------|
| | | D PACKAGE (D16C) | | J PACKAGE (J16A) | | N PACKAGE (N16A) |
| 8-bit | $-55°C \leq T_A \leq +125°C$ | DAC0808LD | MC1508L8 | | | |
| 8-bit | $0°C \leq T_A \leq +75°C$ | | | DAC0808LCJ | MC1408L8 | DAC0808LCN | MC1408P8 |
| 7-bit | $0°C \leq T_A \leq +75°C$ | | | DAC0807LCJ | MC1408L7 | DAC0807LCN | MC1408P7 |
| 6-bit | $0°C \leq T_A \leq +75°C$ | | | DAC0806LCJ | MC1408L6 | DAC0806LCN | MC1408P6 |

*Note. Devices may be ordered by using either order number.

**Figure 3.36** DAC0808 IC device. (*Source:* Reprinted with permission of National Semiconductor Corporation.)

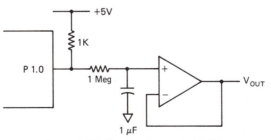

**Figure 3.37** Simple, slow-speed DAC.

## EXAMPLE 3.25 PULSE WIDTH MODULATOR

In the code here, note that timer 0 is being used in mode 2. Also note the push instruction is PUSH ACC instead of PUSH A. The push and pop instructions refer to 8-bit addresses because the 8051 uses on-board RAM for the special function register (SFR) locations.

```
              NUM   EQU    55H        ; DUMMY DATA FOR OUTPUT
              MAX   EQU    0FFH       ; MAX VALUE FOR OUTPUT DATA
              TN    EQU    100        ; TIMER COUNT TO DIVIDE BY 100

                    ORG    0000       ; START HERE ON POWER-UP
    2100            AJMP   MAIN       ; GO TO MAIN

                    ORG    000BH      ; TIMER 0  VECTOR ADDRESS
    8043            SJMP   ISR        ; ALLOW ROOM FOR OTHER
                                      ; INTERRUPT VECTORS
                    ORG    50H        ; ISR STARTS HERE
    C0D0    ISR:    PUSH   PSW        ; SAVE PSW
    C0E0            PUSH   ACC        ; SAVE ACCUMULATOR
    D2D4            SETB   RS1        ; SELECT THE
    D2D3            SETB   RS0        ; REGISTER BANK
    D808            DJNZ   R0,LB1     ; PERIOD FINISHED? JUMP IF NO
    78FF            MOV    R0,#MAX    ; YES, RELOAD PERIOD COUNT
    EA              MOV    A,R2       ; AND RELOAD NUMBER
    F9              MOV    R1,A       ; INTO R1
    D290            SETB   P1.0       ; AND FORCE OUTPUT HIGH
    800B            SJMP   BYE        ; GO TO RETURN
    B90004  LB1:    CJNE   R1,#0,LB2  ; ON-TIME FINISHED? JUMP IF NO
    C290            CLR    P1.0       ; YES, FORCE OUTPUT LOW
    8001            SJMP   BYE        ; GO TO RETURN
    19      LB2:    DEC    R1         ; DEC ON-TIME NUMBER
    D0E0    BYE:    POP    ACC        ; RESTORE ACCUMULATOR
    D0D0            POP    PSW        ; RESTORE FLAGS
    32              RETI              ; RETURN FROM INTERRUPT

                    ORG    100H       ; START OF MAIN
    75A800  MAIN:   MOV    IE,#0      ; DISABLE INTERRUPTS
    C28C            CLR    TR0        ; DISABLE TIMER 0
    758932          MOV    TMOD,#32H  ; INITIALIZE MODE_2
    758C64          MOV    TH0,#TN    ; INITIALIZE TIMER TO
    758A64          MOV    TL0,#TN    ;   DIVIDE BY 100
    D2D4            SETB   RS1        ; SELECT REGS FOR
    D2D3            SETB   RS0        ;   INTERRUPT ROUTINE
    78FF            MOV    R0,#MAX    ; INITIALIZE PULSE PERIOD COUNT
    7955            MOV    R1,#NUM    ; GET DATA INTO R1 FOR ON-TIME
    E9              MOV    A,R1       ;   AND INTO R2
    FA              MOV    R2,A       ;   TO RELOAD R1
    C2D4            CLR    RS1        ; SELECT THE
    C2D3            CLR    RS0        ;   WORKING REGS
    D2AF            SETB   EA         ; ENABLE INTERRUPTS
    D2A9            SETB   ET0        ; ENABLE TIMER 0 INTERRUPT
    D290            SETB   P1.0       ; START HIGH
    D28C            SETB   TR0        ; TURN ON TIMER 1

            ;                 <REST OF PROGRAM>
```

Pay careful attention to the initialization of the IE, TMOD, TCON, TH0, and TL0 registers. Convert the hex values to binary and go back to Chapter 2 for descriptions of what each bit does. For example, the 82H put into IE is 10000010B and activates $\overline{EA}$ and ET0. ET0 high enables the timer 0 interrupt; the other low bits disable the other interrupts.

### 3.12.3 Analog to Digital Converters (ADCs)

A full treatment of A/D design is beyond the scope of this book. In this section we look at the basics of interfacing to an ADC.

There are three main classes of ADCs:

1. *Integrating types* use a technique known as dual-slope conversion. They provide good accuracy at low cost but are slow. They are commonly used in such things as hand-held voltmeters.

2. *Flash converters,* as the name implies, are very fast. Also expensive, they have limited resolution.

3. *Successive-approximation* (SA) types are the garden-variety ADCs. They come in many combinations of speed, accuracy, and cost. The following discussion focuses on SA types.

An ADC consists of two main functional blocks: a *sample-and-hold* (S&H) and the actual A/D. A S&H is needed because an analog signal is typically a "moving target," that is, it is continuously changing while the ADC is trying to convert it to binary. The S&H, in effect, takes a "snapshot" of the analog input and holds it constant while it is converted into digital. The S&H may be built in or on a separate chip.

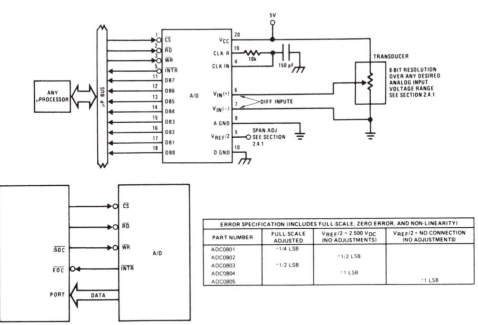

TRI-STATE® is a registered trademark of National Semiconductor Corp.

**Figure 3.38** ADC0801 analog to digital converter interfaced to the 8051. (*Source:* Reprinted with permission of National Semiconductor Corporation.)

The minimum handshake between an ADC and a CPU requires two lines: *start-of-conversion* (SOC) and *end-of-conversion* (EOC). Typically, the program starts the process by pulsing the SOC line, which causes the ADC to take a sample and start the conversion. When the digital word is available, the EOC line is activated. EOC can be polled or used to generate an interrupt. The program responds to EOC by reading the digital word and, if necessary, pulsing SOC to start the process over. Figure 3.39 shows a block diagram and the timing. Note that SOC and EOC are active-low.

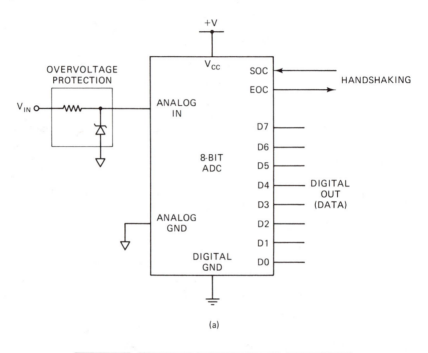

(a)

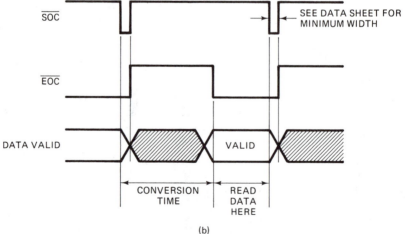

(b)

**Figure 3.39** ADC converter.

## EXAMPLE 3.26 INTERFACING TO AN ADC

```
              SOC     BIT     P2.0      ; DEFINE
              EOC     BIT     P2.1      ;   HANDSHAKE
              BYTE    DATA    P1        ;   AND DATA

                      ORG     40H       ; MAIN STARTS HERE
D2A0   MAIN:          SETB    SOC       ; PULL UP SOC\
D2A1                  SETB    EOC       ; PULL UP EOC\
7590FF                MOV     BYTE,#0FFH ; PULL UP DATA PINS
5100                  ACALL   A2D       ; DO A CONVERSION
00                    NOP               ; REST OF PROGRAM
                                        ;   STARTS HERE

                      ORG     200H      ; ABOVE MAIN
C2A0   A2D:           CLR     SOC       ; ASSERT SOC\
00                    NOP               ; T > MINIMUM WIDTH
D2A0                  SETB    SOC       ; PULL IT BACK UP
20A1FD LP:            JB      EOC,LP    ; WAIT FOR EOC\
E590                  MOV     A,BYTE    ; READ DATA
22                    RET               ; RETURN
```

## 3.13 SUMMARY

In this chapter we discussed mechanical and solid state switches as input devices, including the need for debouncing. We looked at relays and solenoids, including drivers, as output devices. We noted the need for isolation and protection from inductive spikes. *Seven-segment* displays and *multiplexing* were discussed, as were such printing techniques as dot matrix. We defined *handshaking,* and the Centronics, IEEE-488, and RS-232C interfaces were examined. Interfacing applications were done, including stepper motors, printing, and a real-time clock. We discussed the interface between the analog world and the digital microprocessor. D/A and A/D converters were described in terms of the interfacing requirements.

## CHAPTER REVIEW

### Questions

1. Explain contact bounce.
2. What is a limit switch?
3. What does break-before-make mean?
4. Why is hysteresis important?
5. Explain how a switch matrix is scanned.
6. How does two-key rollover differ from N-key rollover?
7. Why is isolation important?
8. Explain what a solenoid is and what it does in a relay.
9. Explain what an inductive spike is and how it is suppressed.
10. Draw a schematic for a 1A1C relay.

11. Compare and contrast a seven-segment and a dot-matrix display.

12. Describe a multiplexed display.

13. What is handshaking?

14. Compare and contrast handshaking on the Centronics and IEEE-488 interfaces.

15. Describe RS-232C.

16. What is a DAC?

17. What is an ADC?

## Problems

1. Write a program to receive serial data. Use a polling loop to detect the start bit.

2. Write a subroutine for the stepper motor of Fig. 3.31 that, when called, will rotate the motor one revolution CCW. Assume 64 steps/rev.

3. Write a program to generate a sawtooth wave by incrementing the accumulator and outputting it to a DAC in a continuous loop. The frequency of the sawtooth wave is determined by calling a time-delay subroutine (kills time) after each output. Assume a 1-MHz clock and get a frequency of 1000 Hz.

4. Rewrite the subroutine of Example 3.25 assuming that both SOC and EOC are active-high.

5. Find the manufacturer's specs for a specific chip, such as an A/D converter, and write a program to use it.

## Chapter 4

# State Machines and Interrupt Timing

**OBJECTIVES**

Upon completion of this chapter, you should be able to

1. Explain what a state machine is and why it is useful

2. Read and make transition diagrams and state tables

3. Explain constraints on interrupt timing

## 4.1 INTRODUCTION

As we pointed out in Chapter 2, when a CPU is combined with ROM, RAM, I/O ports, and other hardware features all on a single chip, the result is a single-chip microcomputer. The most common uses for such devices are as embedded controllers (or bit bangers), which effectively substitute software for hardware. For example, a single 8051 can replace an entire board of TTL chips (including a UART). If we consider that additional chips cost money but additional copies of a program don't, the desirability of replacing hardware with software is obvious. The reduction in hardware made possible by microcontrollers can be seen in Example 4.1.

### EXAMPLE 4.1 AN EMBEDDED CONTROLLER

Figure 4.1 shows a smart modem built with an 80C51, a single-chip modem, and a handful of parts.

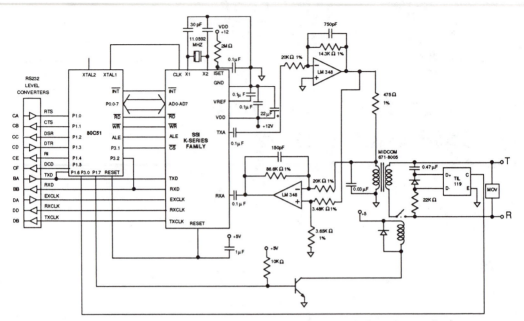

**Figure 4.1** Basic box modem (12V version). (*Source:* Courtesy of Silicon Systems Inc.)

Because of the increasing use of microcontrollers, today's designer of electronic equipment (or electronically controlled equipment) must be as adept at software as at hardware.

## 4.2 FINITE STATE MACHINES

A *finite state machine* (FSM), or just *state machine,* is an abstraction. It can be built with hardware or it can be implemented in code. In software, its purpose is to allow a routine or algorithm to decide what to do next based on both the present set of inputs and the current machine state. State machines are powerful tools for bringing order to complex programming tasks.

At any given time, a program is in one of a finite number of valid *states*. When input arrives (perhaps through an interrupt), the program will make a *transition* to another state. Different inputs will cause different transitions. At each state, some particular action may take place. Before writing the code to implement them, designers put state machines on paper using diagrams, as described below.

### 4.2.1 State Transition Diagrams

An FSM can best be explained by a small example. We will build a *recognizer* for the two words AND and ADD. Assume the words are entered one letter at a time via a keypad. Each time a key is pressed, an interrupt is generated, a letter is read, and the program must decide which of three conditions is true:

Chap. 4: State Machines and Interrupt Timing

1. The word is complete and the appropriate action should be done.

2. More letters are needed to complete the word.

3. An error has been made entering the word.

Note that the full job of the program extends over more than one interrupt from the keypad.

## EXAMPLE 4.2 STATE TRANSITION DIAGRAM

Figure 4.2 shows the state transition diagram for the recognizer routine. We start in state 0 waiting for a key press. Note that in any state, only specific inputs are accepted as valid. Nonvalid intputs will cause an error routine to be executed and the FSM will return to state 0 to start over.

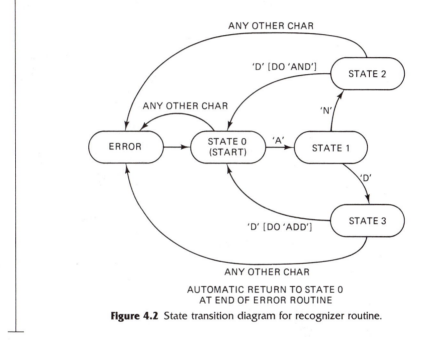

**Figure 4.2** State transition diagram for recognizer routine.

As an example of a nontrivial problem, Fig. 4.3 shows the state diagram for the K224DEMO software package written by Silicon Systems to test its model K224 single-chip modem. The software is written in 8051 assembly language.

### 4.2.2 State Tables

An FSM routine sometimes will keep track of which state it is in by means of a number stored in a RAM location. Other times the state is implicit. When input is received, the routine may decide which transition to make by looking at the state number together with the input. Or, as in Example 4.3, the input will cause different transitions, depending on what state the program is in when the input is received.

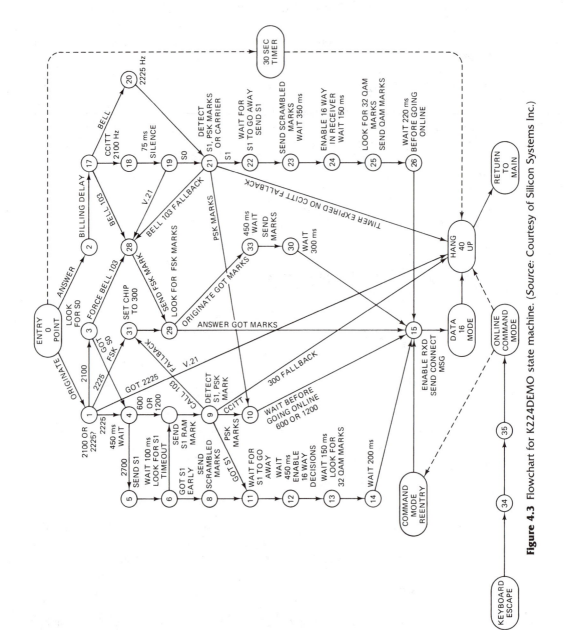

**Figure 4.3** Flowchart for K224DEMO state machine. (*Source:* Courtesy of Silicon Systems Inc.)

An action may be associated with a state. The action taken may be no more than just changing the state number, or it may include some task. An FSM can be summarized in a table. Routines which include the table as a data structure are referred to as *table driven*.

## EXAMPLE 4.3 A STATE TABLE

Figure 4.4 shows the state table for the recognizer.

| | | STATES | | | |
|---|---|---|---|---|---|
| | | 0 | 1 | 2 | 3 |
| INPUTS | 'A' | GO TO STATE_1 | DO ERROR ACTION<br><br>GO TO STATE_0 | DO ERROR ACTION<br><br>GO TO STATE_0 | DO ERROR ACTION<br><br>GO TO STATE_0 |
| | 'N' | DO ERROR ACTION<br><br>GO TO STATE_0 | GO TO STATE_2 | DO ERROR ACTION<br><br>GO TO STATE_0 | DO ERROR ACTION<br><br>GO TO STATE_0 |
| | 'D' | DO ERROR ACTION<br><br>GO TO STATE_0 | GO TO STATE_3 | DO 'AND' ACTION<br><br>GO TO STATE_0 | DO 'ADD' ACTION<br><br>GO TO STATE_0 |
| | ANY OTHER LETTER | DO ERROR ACTION<br><br>GO TO STATE_0 | DO ERROR ACTION<br><br>GO TO STATE_0 | DO ERROR ACTION<br><br>GO TO STATE_0 | DO ERROR ACTION<br><br>GO TO STATE_0 |

**Figure 4.4** State table for the recognizer.

Note that table driven routines can be divided into two sections: the table itself, and the code which uses the table (called the *engine*). Such a division makes it possible to generate a general purpose state machine. All that needs to be changed is a data table. Even if you do not include a table into the program, it is still a good idea to make a table for reference while writing the code.

### 4.2.3 An 8051 Implementation

In this section we develop an 8051 code fragment for the recognizer. The code description is given in Example 4.4.

## EXAMPLE 4.4 RECOGNIZER CODE

The code is given in the accompanying program. Note that we use busy loops to wait in states 0, 1, and 2. We don't have to wait in state 3 because there is no state 4; we just return to state 0. GOTKEY is a bit location and KEY is a byte location, both in RAM. Assume that an interrupt service routine gets a character from a keyboard and puts it into KEY; it then sets GOTKEY high as a flag to indicate a key was pressed.

```
ST_0:    JBC     GOTKEY,TRYA   ; JUMP ON KEY PRESS
         SJMP    ST_0          ; OTHERWISE, WAIT
TRYA:    MOV     A,KEY         ; GET CHARACTER
         CJNE    A,#'A',ERROR  ; IS IT 'A'? GOTO ERROR IF NO

ST_1:    JBC     GOTKEY,TRYN   ; JUMP ON KEY PRESS
         SJMP    ST_1          ; OTHERWISE, WAIT
TRYN:    MOV     A,KEY         ; GET CHARACTER
         CJNE    A,#'N',TRYD1  ; IS IT 'N'?  TRY 'D' IF NO

ST_2:    JBC     GOTKEY,TRYD2  ; WAIT FOR
         SJMP    ST_2          ; NEXT KEY PRESS
         MOV     A,KEY         ; GET CHARACTER
         CJNE    A,#'D',ERROR  ; IS IT 'D'? GOTO ERROR IF NO
         LCALL   DO_AND        ; YES, DO 'AND' ACTION
         SJMP    ST_0          ; BACK TO STATE_0

TRYD1:   CJNE    A,#'D',ERROR  ; IS IT 'D'? GOTO ERROR IF NO

ST_3:    JBC     GOTKEY,TRYD3  ; WAIT FOR
         SJMP    ST_3          ; NEXT CHARACTER
         CJNE    A,#'D',ERROR  ; IS IT 'D'? GOTO ERROR IF NO
         LCALL   DO_ADD        ; YES, DO 'ADD' ACTION
         SJMP    ST_0          ; BACK TO STATE_0

ERROR:   LCALL   DO_ERR        ; DO ERROR ACTION
         SJMP    ST_0          ; BACK TO STATE_0
```

In state 0, the only valid input is 'A'. Receiving 'A' causes a transition to state 1; any other input is an error. When in state 1, the routine will accept as valid input either 'N' or 'D'. The letter 'N' will cause a transition to state 2, and 'D' will cause a transition to state 3. In both state 2 and state 3, the only valid input is a 'D'. The difference is that in state 2 'D' will cause the 'AND' action to be done, whereas in state 3 the same input will cause the 'ADD' action to be done. After either action is complete, the routine returns to state 0.

Note the use of the JBC instruction for testing bits. If the bit is low, JBC will fall through to the next instruction. If the bit is high, JBC will not only jump but will also reset the tested bit back low. Also note that we did not use an explicit state number. The states are implicit in the flow of control.

You could probably come up with an equivalent program without the aid of a state diagram and transition table for the simple case we just examined. But imagine a recognizer for, say, an assembly language. Another example is the K224DEMO software. State machines can keep us from getting totally lost in complicated situations. They can also be automated with the help of special software tools like *scanner generators,* which can create recognizers from a high-level description.

### 4.2.4 FSM Limitations

While state machines are useful tools, they are not the 'universal solvent'. There are many tasks which can not be done by an FSM. A classic example is the job of reading a line of text to determine if it contains an equal number of left and right parentheses. The problem is that an FSM doesn't know what took place in previous states. The test is: if you can't draw a state transition diagram of finite size, you can't make an FSM for the task.

## 4.3 INTERRUPT CONSTRAINTS

Microcontrollers, by definition, are constantly monitoring the state of some external machine or process. However, compared to the instruction execution time of a microprocessor, events in the real world (the external equipment being controlled) often occur very infrequently. Having the processor sit in a polling loop waiting for something to happen ''out there'' would be a waste of time. Therefore, devices such as the 8051 have interrupt schemes allowing for multiple sources, vectoring, and priority. Interrupts are crucial to embedded controller performance and require the examination of three limiting factors: latency, density, and time limit.

### 4.3.1 Latency

*Latency* is the time elapsed from when an interrupt is asserted to when the interrupt service routine begins execution. It includes other factors besides the time required to save the return address and branch. In the case of a single interrupt source, there are only two situations:

1. Interrupts are enabled and the processor is executing an instruction in the main program. The instruction must complete before the interrupt is recognized, so the worst case latency will be increased by the time it takes to execute the longest instruction.
2. The main program is in a critical region where interrupts are disabled. At the end of the critical region the interrupts are reenabled. The worst case latency is then extended by the time it takes to execute the code in the longest critical region.

If there are multiple, prioritized interrupts, the situation is more complicated. In the following discussion we assume that interrupts are disabled when an interrupt service routine starts and are reenabled when it ends; interrupt service routines are not interrupted.

#### EXAMPLE 4.5 A SIMPLE LATENCY CALCULATION

In Fig. 4.5 we want to know the latency for interrupt INT3, which occurs at time t1. The processor cannot respond, however, because it is already executing the service routine for lower priority interrupt INT4, which occurred prior to t1. But before INT4 completes, the higher priority interrupt INT2 occurs, so when INT4 does complete, INT2 starts and INT3 continues to wait. Only when INT2 completes can INT3 finally begin.

The total latency for INT3 includes the time it takes to execute the service routines for INT4 and INT2. In general, the worst case latency time for an interrupt

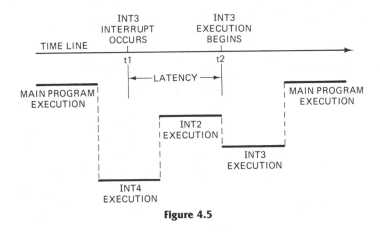

**Figure 4.5**

must include the time to execute the single longest lower priority service routine as well as the time to execute all the higher priority service routines (plus whatever time it takes to get out of the main routine, as discussed above).

Note that the shorter time a high priority interrupt takes to complete the better. Also, the less frequently a high priority interrupt occurs the better, as we shall see in section 4.3.3.

## 4.3.2 Interrupt Density

The basic idea of interrupt density is the percentage of time the processor spends servicing interrupts.

### EXAMPLE 4.6 INTERRUPT DENSITY CALCULATION

If an interrupt occurs once every 10 sec and the service routine takes 2 sec to execute, then the percentage is

$$\text{Density} = (2/10) \times 100\% = 20\%$$

As the percentage approaches 100, the processor gets closer to being *interrupt bound*. When that happens, nothing gets done in the main routine because all processor time is used to service interrupts. The idea can be extended to multiple interrupt sources; see next example.

## EXAMPLE 4.7 DETECTING INTERRUPT-BOUND CONDITION FOR THREE INTERRUPT SOURCES

Suppose INT1 occurs at a frequency of F1 interrupts per second and takes T1 sec to execute its service routine. Likewise, INT2 has a frequency of F2 and takes T2 sec and INT3 occurs at the rate F3 and takes T3 sec. To guarantee that all interrupts have a chance to get serviced, the following inequality should be true:

$$(\text{F1} \times \text{T1} + \text{F2} \times \text{T2} + \text{F3} \times \text{T3}) \times 100\% < 100\%$$

The inequality can be extended to any number of interrupt sources.

### 4.3.3 Interrupt Time Limit

Time and tide wait for no one, nor will most other real-time processes. When an interrupt occurs, the event that caused it may be able to wait only a short time for service before it's too late. For example, imagine a computerized remote probe sent to the surface of a planet. At some point in the descent, the altimeter generates an interrupt to fire the retrorocket. If the processor does not initiate the service routine quickly, the probe becomes scrap metal.

For any interrupt, if the latency time exceeds the maximum allowable waiting time, the system fails (perhaps catastrophically). We have discussed latency already, but now we must combine it with the idea of frequency.

## EXAMPLE 4.8 TOTAL LATENCY TIME

We will modify Example 4.5, as shown in Fig. 4.6. Interrupt INT3 still has to wait for INT4 to finish, and again INT2 occurs before INT4 finishes and forces INT3 to wait more. But now the higher priority INT1 occurs before INT2 finishes and while that's executing, INT2 occurs again. INT3 must wait for INT4, INT1, and two occurences of INT2. Worst case, the total latency time (TLT) for INT3 would be

$$\text{TLT3} = \text{N1} \times \text{T1} + \text{N2} \times \text{T2} + \text{T4}$$

where N1 is the number of times INT1 can occur and N2 is the number of times INT2 can occur before returning to the main program. As before, T$i$ is the time for the INT$i$ service routine to execute.

To see if INT3 really has a chance, we will use TLT3 as follows. Let the period of time between INT3 interrupts be P3, where P3 = 1/F3, and F3 is the interrupts per second rate of INT3. Then, for INT3 to have a chance of getting service, the following must be true: TLT3 $<$ P3. Also, if TMAX3 is as long as INT3 can wait before disaster, then TLT3 $<$ TMAX3 must also be true. The calculation of TLT can be extended to as many interrupts as required, and the inequalities can be generalized to

$$\text{TLT}i < \text{P}i \text{ and } \text{TLT}i < \text{TMAX}i$$

Note that if Tn is the longest of the interrupt service routines of lower priority than INT1, then TLT1 = Tn. In other words, the highest priority interrupt has the best chance of

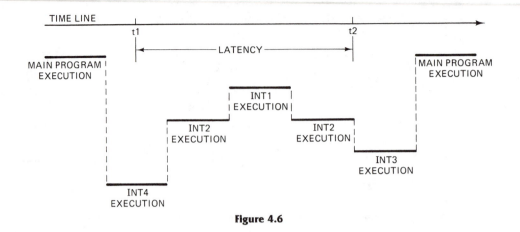

TIME LINE

t1

LATENCY

t2

MAIN PROGRAM
EXECUTION

MAIN PROGRAM
EXECUTION

INT1
EXECUTION

INT2
EXECUTION

INT2
EXECUTION

INT3
EXECUTION

INT4
EXECUTION

**Figure 4.6**

getting service quickly. INT2 has the second-best chance (TLT2 = N1 × T1 + Tn), and so on.

## EXAMPLE 4.9 INTERRUPT TIME LIMIT CALCULATION

Referring to Fig. 4.6, suppose INT3 occurs once per second, and it must be serviced within 400 msec. Assume that we have calculated the execution times of the other interrupts to be T1 = 25 msec, T2 = 50 msec, and T4 = 100 msec. Also assume that, by observing their frequency of occurrence, we know that N1 = 4 and N2 = 2. We calculate as follows:

$$\text{TLT3} = 4 \times 25 \text{ msec} + 2 \times 50 \text{ msec} + 100 \text{ msec} = 300 \text{ msec}$$

F3 = 1/sec, so P3 = 1/F3 = 1000 msec; 300 msec < 1000 msec, so TLT3 < P3 is true; and 300 msec < 400 msec, so TLT3 < TMAX3 is also true. We conclude that INT3 will receive service in time.

For a more detailed mathematical analysis of interrupt timing, see Chapter 3 of *Design with Microcontrollers* by Peatman (McGraw-Hill).

## 4.4 SUMMARY

In this chapter we took a brief look at two issues in using microcontrollers. The first was the idea of a *finite state machine,* which can be a useful concept in organizing certain types of programming tasks. The second issue we examined was the limitations on interrupt driven systems imposed by time constraints, especially when there are more than one source of interrupts. Both are important topics, and the reader is encouraged to do further reading in the literature.

# CHAPTER REVIEW

## Questions

1. What is meant by the state of a process (program)?
2. Use a simple example, other than the ones in this chapter, to explain what is meant by a finite state machine.
3. Explain the time limit problem in an interrupt driven system.
4. What does "interrupt density" refer to?

## Problems

1. Draw a state transition diagram for a recognizer that can distinguish between the words 'CAT', 'CAN' and 'BAT'.
2. Design the state table for problem 1.
3. Write an 8051 assembly language program to implement the recognizer of problem 1.
4. A certain machine control has three push-button switches: forward, stop, and reverse. If the machine is running forward, it must be stopped before it can be run in reverse. Likewise, if it is running in reverse, it must be stopped before running forward. Draw the state transition diagram, draw a state table, and write the 8051 code to implement such a controller.
5. Following Example 4.5, assume a system has three prioritized interrupts: INT1, INT2, and INT3. INT1 has the highest priority and INT3 the lowest. INT1 takes 500 μsec and INT3 takes 300 μsec. Calculate the worst case latency time for INT2, assuming no multiple occurrences of INT3.
6. Calculate interrupt density as a percentage for an interrupt that occurs five times a second and takes 20 msec to execute.
7. A system is planned to have four interrupts with the following specifications:

| Interrupt | Time to execute (μsec) | Frequency of occurrence (times/sec) |
|-----------|------------------------|-------------------------------------|
| INT1      | 500                    | 100                                 |
| INT2      | 800                    | 200                                 |
| INT3      | 400                    | 50                                  |
| INT4      | 100                    | 10                                  |

Calculate whether or not the system will be interrupt bound.

8. Assume a four-interrupt system with the same time to execute numbers as given in Prob. 3. Calculate the total latency time for INT3 (TLT3) if INT1 can happen twice and INT2 can happen three times before INT3 gets its chance.
9. Repeat the problem in Example 4.9 with the following changes: T4 = 50 msec, N1 = 3, and INT3 occurs twice per second.

## Chapter 5

# System Design Techniques

Upon completion of this chapter you should be able to:

1. Describe the steps in the system design cycle.
2. Given a description of a simple electronic design project, write:
   A. System Requirements
   B. System Specifications
   C. Software Modularization
   D. Pseudocode
   E. Assembly Language

## 5.1 INTRODUCTION: THE SYSTEM DESIGN CYCLE

When microprocessors were first used as the basis for electronic systems, a large system might have only a thousand lines of code, but in the past few years microcomputer-based systems have become much more complex. Large systems now might require hundreds of times as much code and contain several microprocessors. The use of *bottom-up* design techniques, where the parts of the system are designed as the need for them becomes apparent, is no longer sufficient. Rather, systematic techniques are called for. This section describes the basic steps used in *top-down* design.

### 5.1.1 System Requirements

The first step in the design cycle is to generate the *system requirements*. This is a document that describes exactly the capabilities desired of the finished system. It must contain a list of all the tasks that the system is to support for the end user. The system requirements are best done in consultation between the design team and the customer or end user of the system.

### 5.1.2 System Specifications

From the system requirements, the *system specifications* are produced. The design team must decide how the system is to interact with the user and specify the inputs and outputs of the device. This second step in the cycle also includes a list of functions that the system must perform in order to implement the user-defined tasks. Whereas the system requirements show what the device must do, the system specifications detail how the system must operate in order to perform the tasks.

### 5.1.3 Partitioning

The third step is to determine how the functions should be *partitioned* between hardware and software. In many cases, it is obvious whether a particular function should be implemented in hardware or software. If there is a possibility of doing it either way, considerations such as cost or development time should point to the more practical method. The general problem of which way to do it is called the *hardware-software trade-off*.

### 5.1.4 Modularization

At this point in the design cycle, the project is split between a hardware design team and a software design team. Each team performs the fourth step, the *modularization*. The hardware team resolves the hardware design into a number of modules, each of which implements one or more of the hardware functions. The software is also partitioned into a hierarchy of modules, each of which performs one or more of the tasks to be implemented in software. The modules might be further divided into procedures, which are routines to perform particular functions. The final software modules consist of groups of procedures that are closely related to each other.

### 5.1.5 Software Design

In this discussion, concentrate on the software side of the cycle. The fifth step is to produce a *pseudo-code* image of the software. Each procedure is written in a syntax-free high-level language. This language has typical high-level program constructs, such as DO WHILE or IF THEN ELSE, and is used to establish the outline of the program, much as a flowchart would be used. Pseudo-code also introduces documentation constructs, to improve the readability and maintainability of the software.

The sixth step in the software design cycle is to verify that the pseudo-code is correct before the software is coded in the actual microprocessor language. A member of the software team other than the writer of the procedure executes the procedure ''in his head'' to make sure that the procedure actually does the function. This step is called the pseudo-code *walk-through*.

The procedure is then coded in some high-level language such as PASCAL or PL/M or in the assembly language for the processor. Each procedure is translated to machine code and tested. This seventh step in the software design cycle results in a set of machine language procedures that have been tested so far as can be in a stand-alone, one-at-a-time mode.

### 5.1.6 Integration and Evaluation

The eighth step involves integrating the software procedures. That is, the procedures are combined and tested for their interaction with each other. This usually involves more than just linking the programs and executing the resulting program. Instead, the highest level procedures (in terms of subroutine calling) are first tested with simplified versions (commonly called *stubs*) of the lower level procedures. As the program flow is verified, more and more complete procedures are substituted until the entire program is tested. If lower level procedures are finished first, they can be tested with short *driver* programs that call them as a higher level procedure would. Tools such as *simulators* and *in-circuit emulators* are available to aid in this task.

The ninth step is to integrate the hardware and software. It might be possible to do this by installing a prom with the completed system software into the hardware and exercising the resultant system. If extensive debugging is required, tools like the microcomputer *analyzer* and in-circuit emulator may be required.

The tenth step is to evaluate the entire system for satisfaction of the original requirements. The customer or end user of the system is the one who ultimately will decide on the suitability of the final product. After all, the customer is the one who must be happy with the performance of the system.

The entire design process is iterative; detection of errors at any point means looping back to an earlier point in the cycle. Figure 5.1 shows a diagram of the software system design cycle.

### 5.1.7 Documentation and Maintenance

In the previous steps, the designers have generated records of all the tasks they have performed during the design cycle. The complete collection of these records, organized in a logical manner, is the *system documentation*. If the records are incomplete or unclear, particularly in a large or complex system, it will be difficult to maintain the software. After the delivery of the system to the customer, it often needs to be modified, either because of undetected errors that only surface in the field or because the system is updated to perform additional functions. This *software maintenance* can be very costly, often accounting for up to 80 percent of the total software costs. Good design techniques and good documentation lead to easier and less costly system maintenance.

### 5.1.8 A Design Example

Let's illustrate the design process by an example: the design of a trainer microcomputer. This is the sort of basic computer that would be used by a student who is learning the assembly language for some processor. The trial programs would be hand-coded into

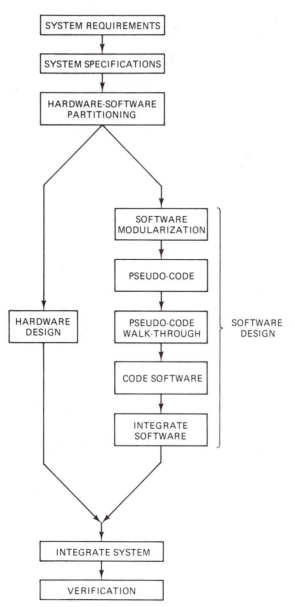

**Figure 5.1** Software design cycle.

machine language, entered, and executed. By this process, the student gets practical experience in writing and debugging software. Because the student is also learning about peripheral devices, the computer would be more useful if it had user I/O ports so that it could interact with and control external hardware. The design example will result in a real product that can be constructed as a project. Appendix A includes a description of the hardware and of the subsystem that controls the displays and keyboard.

## 5.2 TOP-DOWN VS. BOTTOM-UP DESIGN

When a system is designed, it is often immediately apparent that some subsystems are necessary. These subsystems are usually low-level parts of the project that implement very specific tasks. For example, many systems need a keyboard and displays for entry and display of data. If we start by designing these subsystems first, we are embarking on a *bottom-up* design cycle. This might be a reasonable thing to do for a small system that has a small number of subsystems and minimal interaction between them. The bottom-up method allows the designers and programmers to concentrate on the subsystems without having to get bogged down in the planning of the overall system. The problem often comes near the end of the project, when it is discovered that the subsystems do not mesh well, and extensive modification of the subsystems is required. It also might lead to more complicated higher level functions, as the high-level modules bind the subsystems together.

*Top-down* structured design is meant to remedy some of the drawbacks of the bottom-up design. First, the complete system function is specified, and the design is partitioned into a series of subfunctions, each of which might also have subfunctions. In this manner, the complex design is resolved into a series of more easily implemented functions. As each subfunction is specified, it limits and defines the properties of its subfunctions. In this way, the properties of the low-level functions can be expected to combine properly at system integration time.

The design proceeds from consideration of higher to lower modules, from more general to more specific considerations. The highest level is the *system design specification*. For the hardware, the highest level would be a *block diagram;* for the software, it would be the *software modularization*. The lower level functions, such as keyboards and displays, would only be specified later, when it is clear how they will fit in with the higher level modules. Implementation of the low-level functions would not be attempted until the partitioning process was completed.

## 5.3 SYSTEM REQUIREMENTS FOR THE PROJECT

As an example, let us consider the design of a trainer microcomputer. The question is, what should the computer be able to do for the student? A complete and correct answer to this question will lead to the system requirements. There is always the temptation to add too many bells and whistles, but the requirements should include only items actually needed by the user. We will attempt to answer the question in a simple and straightforward way.

### 5.3.1 The Trainer

The unit, like any system, must "come to life" when powered up so that the user knows it is ready to be used. The system has to be ready to accept a user program. Because it is extremely easy to make a mistake when entering a machine language program, it must be feasible to check the program before it is executed. There has to be some way to run the program, and there should be some indication to the user that the execution is proceeding. In case the program does not execute properly, it is important to provide at least some minimal debugging capability. There should be provision for display of an error message if the user makes an operational mistake such as an illegal series of keystrokes. The computer must be able to recover from such an error.

Because the student will be learning about interfacing to peripherals, the unit should have some way to exchange electrical signals with other devices. Finally, because handling devices in real time is an important application to learn about, we will provide the system with the ability to support interrupts.

### 5.3.2 System Requirements

From this discussion we will prepare the *system requirements*. The computer must provide

1. Visual feedback to the user
   a. Show prompt sign on power up
   b. Show error message on illegal operation
   c. Show run indication while executing user programs
2. Ability to enter, examine, and execute a user program. In order to do this, the computer must be able to
   a. Accept any address
   b. Examine the contents of any address
   c. Modify the contents of some addresses
   d. Execute a program from any starting address
3. Program debugging tools
   a. Breakpoint
   b. Display of the CPU register contents
4. Serial and parallel I/O ports
5. Support for interrupts

Many other requirements might be included, such as performance of a self-test upon power up, or having the device prompt the user by issuing instructions about how to operate the system. We also could have included a single-step mode of operation, or specified more complicated I/O, like an interface to a tape drive. We will assume, however, that the list details all the properties of the trainer.

## 5.4 SYSTEM SPECIFICATIONS FOR THE PROJECT

From the list of system requirements, we prepare a detailed description of how the device should interact with its environment and what processes it must perform to satisfy the list. Notice that the requirements describe what tasks the scale should perform. Now we are going to detail how the computer must perform to do the tasks. We will prepare a list of inputs, outputs, and processes to complete the *system specifications*.

### 5.4.1 Inputs

1. Sixteen data keys; a hexadecimal keypad
2. A "high address" key (H)
3. A "low address" key (L)
4. A "store or see" key (S)
5. An "execute" key (G), for "Go"
6. A "reset" key (R)
7. Hardware inputs

### 5.4.2 Outputs

1. Six, seven-segment displays
2. Hardware outputs

The keyboard and display for the computer is shown in Fig. 5.2. The keys labeled U1, U2, and U3 are *user* keys, recognized only by user programs.

### 5.4.3 Processes

1. Upon power up, the five leftmost displays will show the power-on prompt message "ready". The user must then strike any key except the reset key. See Fig. 5.3.
2. Upon the first keystroke, the four leftmost displays will show the starting address 0400, and the other two will show the hexadecimal value of the contents of that address (the data). The computer will wait for the user to strike a key. See Fig. 5.4.

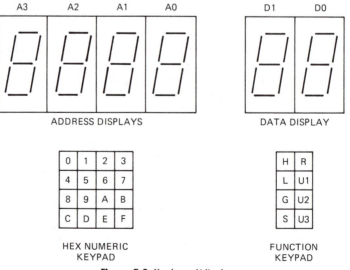

**Figure 5.2** Keyboard/display.

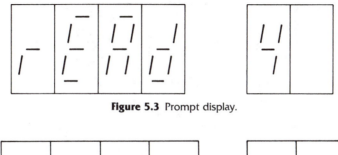

**Figure 5.3** Prompt display.

**Figure 5.4** Starting address display.

3. Striking the R key will cause the same action as power up, independent of the present mode of operation.

4. The H key will be used to modify the address. Depression of H causes the values on the data displays to appear in the two leftmost displays. As the new address appears, the data displays are updated to the contents of that new address.

5. The L key causes the same action as the H key, but it operates on the low byte of the address.

6. Data key depression will cause the data displays to be changed, in a right-entry mode of operation. The rightmost display will accept the key value, and its value will be shifted left to the other data display.

7. The S key allows the user to examine or change the contents of memory. Hitting the S key causes the data shown on the displays to be stored in the memory location pointed to by the current address. The address is then automatically incremented, and the new data are displayed.

8. The G key causes execution of the program stored at the current address. While the program is running, the displays will show ''run.'' See Fig. 5.5.

9. The debugging mode is entered through software. The user must enter a LCALL code into the program at the point she or he wants the program to break. (The user must

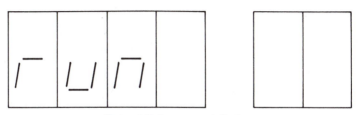

**Figure 5.5** Run prompt display.

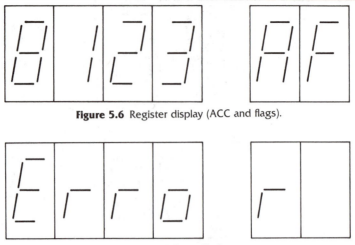

**Figure 5.6** Register display (ACC and flags).

**Figure 5.7** Error display.

replace the proper code later.) When the processor executes the LCALL, the register display mode is entered. The two rightmost displays now identify the register and the contents of that register appear in the four leftmost displays. Any keystroke (except the R key) now changes the register displayed, in the order desired. One more keystroke causes the address of the breakpoint and the LCALL code to show. At this time any key could be used. The contents of the internal registers are not modified by this routine. See Fig. 5.6.

10. "Error" will be displayed if two or more keys are touched at once. See Fig. 5.7. To recover, the operator presses the R key.

11. Serial user I/O is provided by use of the appropriate 8051 pins. Parallel hardware I/O will be done through 8051 ports P0, P1, P2.

### 5.4.4 Comparison

If we compare the specifications with the requirements, it can be seen that the specifications follow from the requirements but are much more detailed. They describe every action that the computer must take in order to perform the user-required tasks. Notice that the specifications are still mostly in terms of English descriptions, with few references to programs or specific hardware. Although specifications 9 and 11 could have been made in more general terms, it has already been decided to use the 8051 microprocessor, and the specifications mirror this decision.

## 5.5 HARDWARE DESIGN

At this point in the design cycle, a decision must be made about whether to implement the various functions in hardware or software. The system needs a microprocessor, ROM and RAM to store the operating software, RAM to store user programs, and interfaces to the keyboard and displays. Figure 5.8 is an example of how the system might be configured.

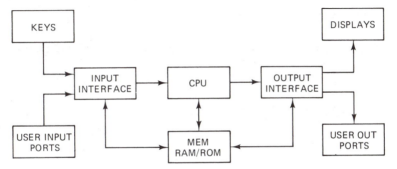

**Figure 5.8** Hardware modular block diagram.

The only blocks in the diagram that might be implemented in software are the blocks for the input and output interfaces. In fact, they are good examples of how the design of a module might be shifted between hardware and software implementation. The keyboard might just be a set of momentary contact keys wired in a row-column matrix. This would be minimal hardware, but the software would be relatively complex. It must scan the matrix, check for key bounces, make sure the same keystroke is not detected more than once, decode keys for their numeric value, and a host of other considerations. On the other hand, the hardware could detect, debounce, decode, and latch the keycodes and provide a key-detected status bit. Then the software would be simple; it would only have to poll the status and input the keycode when it is ready. Similar considerations apply to the output interface, where the displays could be multiplexed in software or hardware.

### 5.5.1 Keyboard/Display Subsystem

Because this chapter is mainly concerned with the software design cycle, the hardware design considerations and descriptions are given in Appendix A. The I/O interface design will be done as a subsystem, and is also presented in the same appendix as another design example. We must, however, specify how the I/O driver will interact with the operating software we are designing.

The interaction is through eight reserved memory locations. Two of these are dedicated to the keyboard. The location KEY_STB is set to one when a legal keystroke is detected, and the location KEYCODE is loaded with the keycode. All that the software we are considering needs to do is poll KEY_STB until it is set, clear KEY_STB, and read KEYCODE. The keycodes are the actual numeric values for the hex keys, and the keycodes for H, L, S, and G are 10H through 13H, respectively. The R key is hardwired to the 8051 reset terminal.

The other reserved locations are six contiguous bytes called DIS_BUF, DIS_BUF+1, . . ., DIS_BUF+5. Our operating system must place the seven-segment codes for whatever we want to display in these locations. The I/O driver multiplexes these codes out to the displays. The code in DIS_BUF is sent to the rightmost display, DIS_BUF+1 to the second rightmost, and so on. All the software has to do to update the displays is write into this display buffer. The seven-segment codes are the same as described elsewhere in this book.

## 5.6 SOFTWARE DESIGN

Many people first learn about programming with a highly interactive language such as BASIC. The tendency is to sit at the terminal and bang out code without sufficient planning. The resulting programs are usually inefficient, hard to document and test, and extremely difficult to maintain. Just as top-down techniques are desirable for system design, so are they desirable for software design. The software design should be *modular,* that is, divided into small, manageable units. *Structured programming techniques,* where all the control and branching operations are done in a standard way, should be used. A plan for the overall program flow must be developed, often with the use of state transition diagrams. The software must be modularized, and pseudo-code written and tested, before the final code for each module can be produced.

### 5.6.1 Top-Down Software Design

Let us consider what a typical top-down structured program looks like. There is a *hierarchical* arrangement of all the routines that make up the package. At the top level is a main driver routine, often called the *executive program.* The function of the executive is to control the overall flow of the program. The actual tasks that must be executed to make the system work are written as a series of subroutines or procedures. The executive contains a section that initializes the system. Then it enters an endless loop that examines the inputs or controls of the system (often through one of the subroutines). The values gathered are tested and the executive conditionally calls the appropriate routines to do the current task. In a more complicated system, the executive might call one of several subexecutive programs, and many of the working procedures might have subroutines of their own. In any case, program control flows from the executive to the other procedures and *always* returns back to the executive.

### 5.6.2 State Transition Diagrams

We must now examine the system specifications to determine what software we need to write. Notice that although there are 21 different input keys, there are only 5 different types of keys. The system behaves in the same way if any of the hexadecimal keys are depressed. We will call them data keys. Striking any of these keys causes the data display to change but does not modify the memory. Both the H and L keys cause the address to change and the data displays to show the new data. H and L will be called address keys. Similarly, G, S, and R have well-defined effects on the state of the system. (The R key is actually implemented in hardware; it is the 8051 reset.)

If we examine the outputs of the system, we can readily group them into address displays and data displays. Perhaps not so obvious is that the state of the system not only is determined by the status of the displays but also by the memory. The H, L, R, and S keys all have the effect of causing the data displays to assume the value of the contents of the current address. Data keys alter this correspondence, and the S key forces the data display value into the memory. The G key passes control away from the program. Thus there are many identifiable output states, depending on the address displays, the data

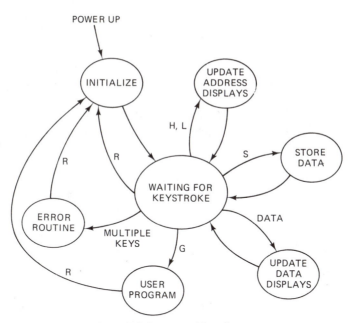

POWER UP

INITIALIZE

UPDATE
ADDRESS
DISPLAYS

H, L

S

STORE
DATA

R

R

WAITING FOR
KEYSTROKE

DATA

ERROR
ROUTINE

MULTIPLE
KEYS

G

R

USER
PROGRAM

UPDATE
DATA
DISPLAYS

**Figure 5.9** State transition diagram.

displays, the memory, and whether we are in a user program. A complete state transition diagram is beyond the scope of this text, but see Fig. 5.9 for a simplified version.

## 5.7 SOFTWARE MODULARIZATION

Once the different procedures or subroutines needed are identified, it is usual to collect them into related groups called modules. There are many reasons for this. For example, all the programs in a particular module might do similar tasks so that they are functionally related. The programs might have to be executed in a narrow time interval, like the procedures invoked by a particular interrupt device. They might be a routine and all its subroutines, or they might just be the programs that didn't seem to fit into any of the other modules.

When we examine the system specifications or the transition diagrams, we see that several programs are going to be needed. In any case, we will need an executive program to control the overall program flow. We will call the executive program MONITOR and define an EXECUTIVE module to hold this. The software to interact with the inputs (through KEY_STB and KEYCODE) will be named GETKEY and will be placed in the INPUT module. Software must be provided to send output codes to the memory locations DIS_BUF, . . ., DIS_BUF+5. These are routines to update the address and data displays. The routines DIS_HI and DIS_LO will be used to modify the contents of the high and low bytes of the address displays. The routine DIS_DAT will change the data displays. Also needed is a routine to cause messages such as ''run'' and ''error'' to appear on the displays. Because these routines will copy set tables of seven-segment codes into the display buffer, we will conceive a program BLK_MOV to do this task. DIS_HI, DIS_LO, DIS_DAT, and BLK_MOV make up the OUTPUT module.

Now we examine the processes section of the specifications. From process A, it can be seen that we need to initialize the address displays and data displays. The hardware for the system will be initialized in that subsystem software. Most systems would have an INIT module, containing programs like INITSW and INITHW to initialize the software and hardware. In our design, we place the procedure INITSW into the EXECUTIVE module. Process B, as mentioned before, is implemented in hardware. Both processes C and D have to do with modifying the address. Module ADDRESS will contain the programs HI and LO, which will be invoked upon detection of an H or L key. Processes E and F will define another module DATA, containing a program DATA to handle data keystrokes and a program STORE. Following specification G, a module RUN will be set up. RUN contains the routine GO, which enters the user program, and the program DIS_RUN. The next two processes suggest a module DEBUG, which contains the programs REGISTERS and DIS_REG and the module ERROR containing ERROR and DIS_ERR. Upon consideration of how the routines might work, we can determine what other routines are needed.

Notice that the routines HI and LO are very similar in function. Each must read the current data display, place the data in the high or low byte of the address displays, and display the new data at that address. Implicit in this process is that the system have a digital image of the current address and data (the image in the display buffer is in terms of seven-segment codes). Assume that the current address is stored in two contiguous memory locations, ADD_BUF and ADD_BUF+1, and the data are stored in location DAT_BUF. The algorithm for implementing HI would be to read DAT_BUF and write those data into ADD_BUF+1 to get the new address. Then that memory location would be read and the result written into DAT_BUF.

All that remains to be done is to update the displays to reflect the new status. The routines DIS_HI and DIS_DAT could be called to do this. But because these programs read a digital image from the data and address buffers, and must place a seven-segment code image in the display buffer, they might each call a routine called CVT. CVT would be a look-up routine that is entered with a hexadecimal digit and returns the seven-segment code. The CVT routine accesses a table of seven-segment codes that is stored in the system ROM.

In addition to CVT, this discussion introduces two other programs. We define the procedure STORAGE1, which sets up reserved memory locations for the address and data buffers and any others we might need. The procedure MES_TBL will define the seven-segment code table and any other tables we must implement.

Consideration of how STORE and DATA work does not lead to any more new procedures. The programs DIS_ERR and DIS_RUN need seven-segment code tables for their messages, but these can be defined in procedure MES_TBL. The DISREG routine needs reserved memory to save the registers; this can be done in STORAGE1.

This completes the software modularization. There are, of course, a great variety of ways the procedures and modules can be chosen. For example, DIS_RUN, DIS_PMT, and DIS_ERR cause a message to show on the display; they could have been grouped together in a module called MESSAGE. A diagrammatic representation of the software modules is shown in Fig. 5.10.

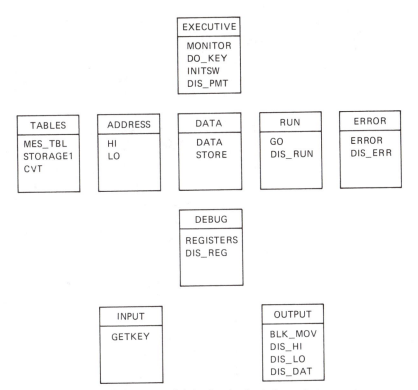

**Figure 5.10** Software modularization for the trainer microcomputer.

At this point it is usual to introduce another diagram, the procedure calling tree, shown in Fig. 5.11. This diagram organizes the procedures in a way that highlights the relationship between called and calling procedures. The diagram includes a notation to specify parameters passed between the procedures.

## 5.8 PSEUDO-CODE

Nearly everyone who learns how to program is introduced to the technique of using flowcharts to aid in the program development. Flowcharts, which are graphical devices to depict the algorithm for the software being constructed, have the advantage of familiarity. However, they have several disadvantages. Unlike pseudo-code, the flowchart does not force the programmer to follow tenets of modular, structured programming. There is seldom room on the flowchart to describe program steps fully. Furthermore, it is difficult to keep the documentation current; any change involves a complete redrawing of the chart.

Pseudo-code, on the other hand, contains documentation and program structures and is easily updated with a word processor. Pseudo-code, or programming design language (PDL), is the technique we will describe and employ.

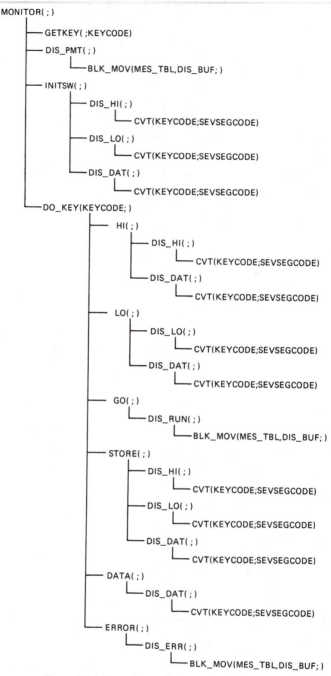

**Figure 5.11** Procedure calling tree for the trainer.

Chap. 5: System Design Techniques

What are the properties that a pseudo-code should possess? The code should, like such languages as PASCAL or PL/M, support the constructs necessary to impose proper structure to the software. Such languages, however, are meant to be read by computers and are very inflexible with respect to syntax. The pseudo-code should not need exact punctuation or have a special set of reserved words. Rather, the ideas that express the algorithm for the software should be coded in plain English statements. For example, if we wanted to set a counter to zero, the PASCAL statement would be "COUNT :=0;". In pseudo-code, it could be "SET COUNT TO ZERO," or "RESET COUNT," or any other statement that is easily understood by the reader.

Documentation constructs that encourage the programmer to use modular design and to provide a complete description of the software are an important part of the design language. This includes items such as packages to outline modules and procedures and headers, or prologues, to document properly the contents of the package. Each module could be declared a MODULE . . . END MODULE construct. The module prologue might contain such items as authorship, date, a list of the procedures contained, and a list of data accessed or modified. The procedure declaration might be similar, such as PROCEDURE . . . (header) . . . BEGIN PROCEDURE . . . (code) . . . END PROCEDURE.

Program constructs must be included to ensure proper structured design. Each construct defines a program segment that has single entry and exit points and implements a certain, well-defined function. It is possible to define a set of program constructs such that proper program flow control is easily attained. The structures could be classified as

1. Sequence of statements

2. Conditional structures

3. Loop constructs

The sequence construct would be "DO (series of statements) END". It is possible to invoke an entire procedure (which could be thought of as a sequence) with the "CALL . . . RETURN" construct. The conditional structures would be "IF(condition)THEN(series of statements)" and "IF(condition)THEN(series of statements) ELSE (series of statements)." The loop constructs would include the endless loop "DO ALWAYS . . . END" and conditional loops "DO WHILE (condition). . .END" and "DO UNTIL(condition) . . . END." Other constructs also follow the rules for good structure, but we will use only those mentioned here. The "GOTO . . ." structure will be avoided.

An important aspect of modular design is how to keep track of parameters. As each procedure is invoked, the calling procedure might be required to send some information to its subroutine. This routine might also want to pass data back to the calling routine. Each procedure name will be appended with an input and output parameter list to keep track of the information flow. The format will be

NAME (parm1; parm2)

where parm1 is an input parameter and parm2 is an output parameter. Multiple parameters will be separated with commas. An example would be an addition routine; the calling procedure would pass in the two addends and receive back the sum. The complete name of such a routine might be ADDER(addend1,addend2;sum).

We will illustrate the use of design language by continuing the design of the trainer microcomputer.

### 5.8.1 Pseudo-Code for the Module Executive

A review of the design specifications, the state transition diagram, the software modularization, and what we have said about how the software driver must work enables us to develop an algorithm for the MONITOR procedure. Evidently, it must invoke DIS_PMT to set up the initial display status. Then it would call GETKEY to wait for the first keystroke. Upon the first keystroke, and after return to MONITOR, INITSW must be called to initialize the buffers and displays to show the starting address. The program would then go into an endless loop, at the head of which is an invocation of the procedure GETKEY to detect key closures by the operator. The procedure DO_KEY would be called in order to test the keycode and selectively invoke the appropriate routine, one for each type of key. Notice that the control always returns to MONITOR, and thus to GETKEY, where the computer is waiting for the next keystroke.

The DO_KEY routine may call six other routines: HI, LO, DATA, GO, STORE, and ERROR. The author of the DO_KEY software does not need to know the details of how each of these routines is constructed. All that is required is the assurance that a routine to implement the task described in the specifications exists or will exist, and what parameters must be passed. We will note that GETKEY gathers the keycode and returns it to MONITOR. Thus, GETKEY has an output parameter KEYCODE. The procedure DO_KEY needs to know which key was pressed; this information is passed into DO_KEY as the input parameter KEYCODE. No other routines directly pass information, but all of the other routines access or modify the memory locations ADD_BUF and DAT_BUF. These locations will be defined as program variables.

### 5.8.2 Pseudo-Code for the Procedure MONITOR

A first pass at writing the pseudo-code for MONITOR is shown in the following segment.

```
PROCEDURE MONITOR

    (PROCEDURE HEADING TO BE ADDED LATER)

BEGIN PROCEDURE MONITOR

        CALL DIS_PMT
        CALL GETKEY
        CALL INITSW
        DO FOREVER
            CALL GETKEY
            CALL DO_KEY
        END

END PROCEDURE MONITOR
```

Examine the code first for the documentation constructs. The prologue is always placed between the procedure declaration and ''BEGIN PROCEDURE,'' and the body, or executable code, follows. Indentions and blank lines set off the different parts of the program to facilitate easy reading. The body of the program is a series of statements; look for some of the program constructs we have defined. Indention is also employed to set off the various program constructs. We have yet to complete the procedure heading and to show the relationship for the parameters.

Now we verify that the code actually implements the task that has been set. Does this code cause the ready prompt to appear upon power up? Does it then wait for the user to touch a key? Is there provision for setting up the proper starting address? Does it finally go into an endless loop that monitors the keypad and invokes the routine that implements the command issued from the keypad? This process of verification by a thought process is the design walk-through and must be done for each procedure before conversion to the final programming language.

### 5.8.3 Pseudo-Code for the Procedure INITSW( ; )

The procedure INITSW should initialize the memory locations ADD_BUF and DAT_BUF. After the address and data displays are updated to reflect this new information, the task of INITSW is over. Let us write the pseudo-code for INITSW, but this time attempt to include a proper procedure heading and keep track of any parameters or variables. The code is shown in the following procedure.

```
PROCEDURE INITSW( ; )

    WRITTEN BY: G. THOMAS HUETTER   DATE: 2/21/88

    CALLED BY:
         DIS_HI( ; ), DIS_LO( ; ), DIS_DAT( ; )

    DATA STRUCTURE:
         INPUT PARAMETERS:    NONE
         OUTPUT PARAMETERS:   NONE
         PROGRAM VARIABLES:   ADDRESS BUFFER, DATA BUFFER

    BEGIN PROCEDURE INITSW( ; )

         SAVE ENVIRONMENT
         SET ADDRESS BUFFER TO 1400H
         CALL DIS_HI( ; )
         CALL DIS_LO( ; )
         READ CONTENTS OF 1400H
         STORE IT IN DATA BUFFER
         CALL DIS_DAT( ; )
         RESTORE ENVIRONMENT
         RETURN

    END PROCEDURE INITSW( ; )
```

INITSW does not accept any parameters from MONITOR, nor does it pass any back. Although there are no parameters in this routine, there must be a standard notation to identify the parameters of a procedure. The notation for this procedure name will be INITSW( ; ). The field inside the parentheses is meant to contain the parameter list, with input parameters preceding the semicolon and output parameters following. Each procedure should have information in the procedure header. This should include authorship, purpose of the program, how the procedure is invoked, and the programs it invokes and should show how the program handles data. The subheading data structure is used to describe any parameters and to show variables or memory locations that are modified by the procedure.

As for the body of procedure INITSW( ; ), we should perform a walk-through to verify correct operation. Does it properly initialize the address buffer and data buffer? Does it cause the proper response on the displays? When answering these questions, we assume that properly functioning subroutines such as DIS_HI( ; ), DIS_LO( ; ), and DIS_DAT( ; ) exist.

### 5.8.4 Pseudo-Code for the Procedure DO_KEY(KEYCODE; )

The code for this procedure must identify which key was struck and invoke the appropriate procedure to implement the function desired for that keystroke. For example, when any data key is touched, the routine DATA should run to cause the value of the data key to appear in the data displays. This routine must know the value of the keycode, but returns no data back to DO_KEY(KEYCODE; ); thus it will be called DATA(KEY-CODE; ). Similarly, strokes of the H, L, G, and S keys must, respectively, call the routines HI( ;), LO( ;), GO( ; ), and STORE( ; ). Any other key should lead to execution of the routine ERROR( ; ). Walk through the pseudo-code shown here to verify proper operation. Notice that the procedure prologue shows authorship, has a thumbnail description of the purpose of the software, and identifies KEYCODE as the only datum involved.

```
        PROCEDURE DO_KEY(KEYCODE; )

            WRITTEN BY:  G. THOMAS HUETTER  DATE: 2/21/88

            CALLED BY:
                 MONITOR( ; )

            PROCEDURES CALLED:
                 HI( ; ), LO( ; ), DATA(KEYCODE; ), STORE( ; )
                 GO( ; ), ERROR( ; )

            DATA STRUCTURE:
                 INPUT PARAMETERS:  KEYCODE
                 OUTPUT PARAMETERS: NONE

            BEGIN PROCEDURE DO_KEY(KEYCODE; )

                 IF KEYCODE IS FOR "H" KEY
                     THEN CALL HI( ; )
                 IF KEYCODE IS FOR A DATA KEY
                     THEN CALL DATA(KEYCODE; )
                 IF KEYCODE IS FOR "L" KEY
                     THEN CALL LO( ; )
                 IF KEYCODE IS FOR "G" KEY
                     THEN CALL GO( ; )
                 IF KEYCODE IS FOR "S" KEY
                     THEN CALL STORE( ; )
                 IF KEYCODE IS AN ILLEGAL KEY
                     THEN CALL ERROR( ; )
                 RETURN

            END PROCEDURE DO_KEY(KEYCODE; )
```

## 5.8.5 Pseudo-Code for the Procedure DIS_PMT( ; )

Upon power up, the computer is supposed to show the prompt "ready" in the displays. In order to do this, a table of seven-segment codes stored in the ROM must be copied to the display buffer. Because there are other messages to be sent out at other times, we have a program BLKMOV that can copy the required 6 bytes from one set of memory locations to another. What DIS_PMT( ; ) must do is to set up pointers to the message table and to the display buffer. The parameters MES_TBL and DIS_BUF are the respective pointers. MES_TBL can have different values, depending on the message to be sent—in this case, PMT_TBL. The copy subroutine will be called BLKMOV(MES_TBL,DIS_BUF; ) and in this instance MES_TBL will be PMT_TBL.

The pseudo-code for DIS_PMT( ; ) is shown below. Verify proper operation and documentation for the procedure. The procedures DIS_RUN( ; ) and DIS_ERR( ; ) would be identical except for the value of MES_TBL.

```
PROCEDURE DIS_PMT( ; )

    WRITTEN BY:  G. THOMAS HUETTER  DATE: 2/20/88

    CALLED BY:
        MONITOR( ; )

    PROCEDURES CALLED:
        BLK_MOV(MES_TBL,DIS_BUF; )

    DATA STRUCTURE:
        INPUT PARAMETERS:  NONE
        OUTPUT PARAMETERS: NONE

    BEGIN PROCEDURE DIS_PMT( ; )

        POINT TO DISPLAY BUFFER
        POINT TO PROMPT MESSAGE TABLE
        CALL BLK_MOV(MES_TBL,DIS_BUF; )
        RETURN

    END PROCEDURE DIS_PMT( ; )
```

## 5.8.6 Complete Pseudo-Code for the Module EXECUTIVE

Following is the complete pseudo-code for the module EXECUTIVE.

```
MODULE EXECUTIVE

    WRITTEN BY:  G. THOMAS HUETTER
    DATE:        3/2/88

    PROCEDURES: MONITOR( ; ), INITSW( ; ), DO_KEY(KEYCODE; ),
                DIS_PMT( ; )

    EXTERNAL PROCEDURES: GETKEY( ; ), DIS_HI( ; ), DIS_LO( ; ),
                         DIS_DAT( ; ), HI( ; ), LO( ; ), GO( ; ),
                         DATA(KEYCODE; ), STORE( ; ), ERROR( ; )
```

```
PROCEDURE MONITOR( ; )

        WRITTEN BY:  G. THOMAS HUETTER  DATE: 2/19/88

        CALLED BY:
             THIS PROCEDURE IS INVOKED UPON POWER-UP
             OR UPON A COLD OR WARM RESTART.

        PROCEDURES CALLED:
             DIS_PMT( ; ), GETKEY( ;KEYCODE),
             INITSW( ; ), DO_KEY(KEYCODE; )

        DATA STRUCTURE:
             INPUT PARAMETERS:  NONE
             OUTPUT PARAMETERS: NONE

        BEGIN PROCEDURE MONITOR( ; )

             CALL DIS_PMT( ; )
             CALL GETKEY( ;KEYCODE)
             CALL INITSW( ; )
             DO FOREVER
                  CALL GETKEY( ;KEYCODE)
                  CALL DO_KEY(KEYCODE; )
             END

        END PROCEDURE MONITOR( ; )

PROCEDURE INITSW( ; )

        WRITTEN BY:  G. THOMAS HUETTER  DATE: 2/21/88

        CALLED BY:
             MONITOR( ; )

        PROCEDURES CALLED:
             DIS_HI( ; ), DIS_LO( ; ), DIS_DAT( ; )

        DATA STRUCTURE:
             INPUT PARAMETERS:  NONE
             OUTPUT PARAMETERS: NONE
             PROGRAM VARIABLES: ADDRESS BUFFER, DATA BUFFER

        BEGIN PROCEDURE INITSW( ; )

             SAVE ENVIRONMENT
             SET ADDRESS BUFFER TO 1000H
             CALL DIS_HI( ; )
             CALL DIS_LO( ; )
             READ CONTENTS OF 1000H
             STORE IT IN DATA BUFFER
             CALL DIS_DAT( ; )
             RESTORE ENVIRONMENT
             RETURN

        END PROCEDURE INITSW( ; )

PROCEDURE DO_KEY(KEYCODE; )

        WRITTEN BY:  G. THOMAS HUETTER  DATE: 2/21/88

        CALLED BY:
             MONITOR( ; )
```

```
                PROCEDURES CALLED:
                    HI( ; ), LO( ; ), DATA(KEYCODE; )
                    STORE( ; ), GO( ; ), ERROR( ; )

                DATA STRUCTURE:
                    INPUT PARAMETERS:  KEYCODE
                    OUTPUT PARAMETERS: NONE

                BEGIN PROCEDURE DO_KEY(KEYCODE; )

                    IF KEYCODE IS FOR "H" KEY
                        THEN CALL HI( ; )
                    IF KEYCODE IS FOR A DATA KEY
                        THEN CALL DATA(KEYCODE; )
                    IF KEYCODE IS FOR "L" KEY
                        THEN CALL LO( ; )
                    IF KEYCODE IS FOR "G" KEY
                        THEN CALL GO( ; )
                    IF KEYCODE IS FOR "S" KEY
                        THEN CALL STORE( ; )
                    IF KEYCODE IS AN ILLEGAL KEY
                        THEN CALL ERROR( ; )
                    RETURN

                END PROCEDURE DO_KEY(KEYCODE; )

            PROCEDURE DIS_PMT( ; )

                WRITTEN BY:  G. THOMAS HUETTER   DATE: 2/21/88

                CALLED BY:
                    MONITOR( ; )

                PROCEDURES CALLED:
                    BLK_MOV(MES_TBL,DIS_BUF; )

                DATA STRUCTURE:
                    INPUT PARAMETERS:  NONE
                    OUTPUT PARAMETERS: NONE

                BEGIN PROCEDURE DIS_PMT( ; )

                    POINT TO DISPLAY BUFFER
                    POINT TO PROMPT MESSAGE TABLE
                    CALL BLK_MOV(MES_TBL,DIS_BUF; )
                    RETURN

                END PROCEDURE DIS_PMT( ; )


        END MODULE EXECUTIVE
```

This document collects together all the procedures for the module EXECUTIVE. The module header shows authorship, a list of the procedures included, and a list of external (i.e., in a different module) procedures called. Walk through each procedure in the module to verify correct operation.

A similar process must be performed for each module in the system. When this is done, a complete pseudo-code implementation of the system software will be available. The completion of the pseudo-code for the microprocessor trainer is left as an exercise.

## 5.9 ASSEMBLY LANGUAGE

The software must now be translated into machine code for the microprocessor being used—in this case, 8051 machine code. However, the pseudo-code is never translated directly into machine code but rather is encoded into some high-level language such as PASCAL, PL/M, or C. A compiler would then generate the machine language for each procedure from the high-level language implementation. Critical sections of the software, for example, procedures that must execute at maximum speed, would be written in the assembly language for the processor.

The task of generating the machine level code has several levels. Each pseudo-code procedure must be converted to 8051 assembly language. Each must then be assembled. The output of the assembler would include an object-code program, which would be machine language, but probably not in executable form. These individual machine code programs must then be combined or linked into a complete executable program. The software must be tested for proper operation, and if there are errors they must be located and corrected. Testing should not wait for completion of the linking procedure; there are techniques for testing individual machine code programs.

### 5.9.1 Assembly Language for the Module EXECUTIVE

It is now time to encode the pseudo-code procedures into assembly language. As examples of this process, we will generate assembly code for one of the four programs in the module EXECUTIVE. Refer to the pseudo-code for the module.

**Assembly language for the procedure MONITOR( ; ).** The MONITOR procedure is entered by the command JMP MONITOR in the program INIT, which is part of the I/O subsystem described in Appendix A. Therefore, the program must have an entry label MONITOR, which must be declared global. The ( ; ) notation must be suppressed in the assembly language, as the assembler will only accept alphanumeric characters (and the character _ ) as components of a symbol. Because MONITOR calls four other routines—GET_KEY, DIS_PMT, DO_KEY, and INITSW—these symbols must be declared external. There are no parameters to be assigned. The body of the procedure involves the first three calls and then an endless loop of alternate calls of GET_KEY and DO_KEY.

Each assembly routine should also have a header. This should include similar information to the pseudo-code header, such as authorship, calling routine(s), called routine(s), and parameters, and it should keep track of any registers or memory locations affected. Another necessary item for the HP64000 assembler is the notation ''8051,'' which tells the system the type of microprocessor for which the assembly code is written. Examine the following segment and notice the relationship between the pseudo-code and assembly code.

```
        "8051"

;THIS IS THE MONITOR PROGRAM
;WRITTEN BY: G. THOMAS HUETTER
; CALLED BY: INVOKED AT POWER-UP, OR UPON EXECUTION OF "RST 7"
; SUBROUTINES CALLED: DIS_PMT, GETKEY, INITSW, DO_KEY
```

```
; PARAMETERS:    INPUT: NONE         OUTPUT: NONE
; REGISTERS AFFECTED: PSW
; MEMORY LOCATIONS AFFECTED: ADD_BUF, ADD_BUF+1, DAT_BUF
;                           DIS_BUF THROUGH DIS_BUF+5
;
          GLB    MONITOR
          EXT    GETKEY, DIS_PMT, DO_KEY, INITSW

          CODE

MONITOR:  CALL   DIS_PMT    ;DISPLAY THE POWER-ON PROMPT
          CALL   GETKEY     ;WAIT FOR FIRST KEYSTROKE
          CALL   INITSW     ;DISPLAY STARTING ADDRESS
L1:       CALL   GETKEY     ;GET THE NEXT KEYCODE
          CALL   DO_KEY
          JMP    L1
```

## 5.9.2 Software Integration

We now test and combine the individual procedures into a complete functioning program. Each assembly language procedure must be assembled, a process that generates an 8051 machine code image of the procedure. However, these routines are usually not executable routines, because of undefined quantities, the variables that have been declared external. The external variables are evaluated by the linker, as soon as the program that has defined the external has been linked to it. It might thus appear that all the procedures must be completed and linked together before any testing may be done. It is better to write simplified versions (called *stubs*) of the lower level procedures, which contain just enough information to define the externals and parameters required for that procedure. Then the link process can proceed properly, and the resulting code can be tested at least to the extent that the completed procedures (which would be the higher level procedures) successfully invoke the yet unwritten lower level programs. A repetitive process of executing the software on a simulator, software emulator, or other debugging tool, fixing any problems that might crop up and adding in completed lower level procedures, will result in a complete machine language of the software. This software will have been tested so far as possible without interaction with the actual system hardware.

As an example, suppose that the software already described in the assembly language section had been written, but not the rest. If these procedures were submitted to the linker, many unresolved external errors would be reported, and the resulting machine code could not execute properly. In the procedure DO_KEY alone, for example, there are six undefined variables—HI, LO, GO, STORE, DATA, and ERROR. Each of these is a subroutine name, and each would be defined temporarily by a stub. Each stub could be as simple as the one shown for HI:

```
          ''8051''

          ;STUB FOR HI

                    GLB    HI

          HI:       RET
```

For routines with input or output parameters, the stub would be somewhat more complicated because it would have to simulate proper operation of the parameters.

Once the software to be tested and its stubs have been linked, the resulting program could be tested. We might single step through the program. We could begin execution from the label MONITOR and verify that all of the routines are invoked at the proper time. Notice that when we call GET_KEY from the label L1 the program is to return with some keycode in the accumulator. This could be done by a simulation in GET_KEY or by using the register modification function of the simulator. Subsequent execution of DO_KEY could then be checked for invocation of the proper routine (HI, LO, etc.). As each routine is completed, it could be added and tested, until the software is complete.

## 5.10 SYSTEM INTEGRATION AND EVALUATION

At this point, the hardware and software must be combined. A possible way to accomplish the task is to burn the software into an EPROM and install the EPROM into the system hardware. The system is then fired up and exercised in order to evaluate the operation. It is easily possible that the software does not operate properly in the real system hardware, even if it seemed to work correctly in the simulations of the software integration phase. This could be caused by either software or hardware problems, and such bugs must be traced and corrected, using tools such as logic analyzers and microcomputer analyzers. Each time a mistake is found in the software, the EPROM must be erased and reburned. In large systems, or in systems where there is a lot of interaction between the hardware and the software, this method is likely to be quite difficult and time-consuming.

### 5.10.1 The In-Circuit Emulator

A tool that can greatly enhance the integration phase is the in-circuit emulator (ICE). This consists of an emulation pod and emulation software. The emulation pod contains a microprocessor that is connected both to the development system and to a ribbon cable with a socket plug. The plug is inserted into the microprocessor socket of the system hardware, so that the processor in the pod becomes the system (called the "target" system) processor. Now, when the target system is exercised, the development system, under the influence of the emulation software, can monitor and control execution of the software under test. A full range of debugging tools is usually available, including memory and register examination and modification, breakpoints, single stepping, program tracing, and control of the I/O ports.

An important feature of the ICE is the existence of emulation memory. This is memory in the development system that can be used to overlay memory in the target system. When the target system is exercised, the overlay memory appears to be part of the target system. This can be very advantageous during integration; for example, the system can be tested before the entire system memory has been constructed. In fact, the entire target system memory map could be overlayed in emulation memory, and the software integration could be performed by use of the ICE. Another important use of the overlay memory is to overlay the system ROM. Then the system software can be loaded by the emulator, eliminating the necessity to burn an EPROM. Each cycle of discovering and eliminating a software bug then means a new load of the program rather than reburning an EPROM.

Because the emulator can control the memory and I/O of the target system, it can also be used for testing the system hardware. Even in the absence of any system software, the target hardware can be exercised through the memory and I/O commands of the ICE and tested with short test programs that simulate some of the functions of the target system. The hardware can also be tested in a modular manner because some parts of the system memory can be provided temporarily by emulation memory. Use of the ICE can yield considerable economy of time and effort in hardware integration and testing.

If the software integration phase has been completed, and the hardware has been tested in a stand-alone mode, then system integration can proceed, except that the need for reburning EPROMS is eliminated and the sophisticated debugging features of the ICE are available.

### 5.10.2 Integration in Concert

Rather than individual integration of the software, hardware, and system, it is often advantageous to proceed with integration of all phases in concert. In order to do this, a hardware prototype must be available early in the software development process. Then the early software integration can be done with the use of an ICE along with the actual system hardware. This allows a modularized, top-down approach to the integration and testing, where the hardware and software are done together.

A typical scheme for in-concert integration would involve preparation of the highest level software module (the EXECUTIVE) and the lowest level ones (the INPUT and OUTPUT). The software in these modules is then linked with any stubs required to resolve the external variables and satisfy the requirements for parameters. Similarly, the hardware should have the highest (CPU) and lowest (I/O) level modules completed. Now the ICE is used to provide emulation memory and the system can be tested for proper operation of the I/O (keyboard and display, for example) and for overall program flow. Problems that crop up can much more easily be found and repaired than if we waited until the entire system was constructed (as at system integration time in the scheme above).

As software and hardware modules are completed, they are integrated into the system. At each step, as the functionality of the system is increased, testing and debugging continue. When all the modules are in place, the system is ready for the final evaluation to make sure that it satisfies all the original requirements and specifications.

## 5.11 SUMMARY

In this chapter we have discussed some techniques for design of microprocessor-based electronic systems. We have described the software system design cycle and have shown the importance of proper top-down design. As an example of these techniques, we have embarked upon the design of a simple single-board computer, such as might be used as an aid in learning assembly language.

The design of this trainer computer was initiated with the generation of the system requirements. From them, the system specifications were written, and a software modularization was constructed. A particular module was then selected (the EXECUTIVE module), and pseudo-code and assembly language were written to implement the module.

Finally, we discussed software integration, system integration, and evaluation of the system.

## REFERENCES

FREEDMAN AND EVANS, *Designing Systems with Microcomputers* (Englewood Cliffs: Prentice Hall, 1983).

YOURDAN, *Structured Walkthroughs* (Englewood Cliffs: Prentice Hall, 1980).

## CHAPTER REVIEW

### Problems

1. Generate the system requirements for an automated scale, such as might be used in a delicatessen or butcher shop. Consider the following questions (and others that you can think of) as an aid:
   - How would the unit determine the weight of the product?
   - How would the user enter the unit cost (price/lb) of the product?
   - How would the scale display the weight, unit cost, and total cost to the user and to the customer?
   - How would the scale compensate for the tare of a container holding the product to be sold?
2. Write system requirements for a point-of-sale (POS) terminal, as would be used in a fast-food restaurant. First list an appropriate set of questions similar to those in Prob. 2.
3. Write system specifications for the automated scale of Prob. 1.
4. Write the specifications for the computer-controlled POS terminal of Prob. 2.
5. Assume that the scale of Prob. 1 uses a load cell and an A/D converter to weigh the item, so that a digital image of the weight is available, and that an I/O driver subsystem similar to the one used for the trainer's keypad and displays exists. Generate a software modularization for the system, including a state transition diagram, a thumbnail description of each procedure, a module diagram, and a procedure calling tree.
6. Write the modularization for the POS terminal.
7. Write pseudo-code for the programs in the module OUTPUT of the trainer. Review the material in the specifications and software modularization sections to see how the procedures work. Include proper documentation, and keep track of any parameters used.

8. Write pseudo-code for the procedures in the modules ADDRESS and STORE.

9. Complete the pseudo-code for the trainer.

10. Write assembly language for the procedures in the module OUTPUT. Write the code in a fashion that a typical assembler would accept (proper syntax). Don't forget global and external declarations and include headers and comments.

11. Write the assembly for the procedures in the modules ADDRESS and DATA.

12. Complete the assembly language for the trainer.

# **Appendices**

# Appendix A

# Project Design

## A.1 HARDWARE DESCRIPTION OF THE PROJECT

### A.1.1 The Chip Set for the Project

An Intel 8051 will be used for the CPU.

### A.1.2 Memory Map

To be designed by student.

### A.1.3 Schematic for the Project

The schematic for the microcomputer trainer project is to be drawn by the student. The connections between the 8051 and the keypad/display are shown on the keypad/display subsystem design, Fig. A.1. The reader should verify that the schematics are consistent with the memory and I/O maps shown on following page.

## A.2 THE I/O DRIVER SUBSYSTEM

This section shows the documentation for the design of the keypad/display driver subsystem. The properties of this subsystem are outlined in Sec. 5.5.1. Table A.1 shows a summary of the reserved memory locations used by the I/O subsystem. Table A.2 shows, for each key, the binary number output on Port C (SEL_PORT), the corresponding number input on Port B (KEY_PORT), and the final keycode that is placed in the location KEYCODE.

**Table A.1**

RESERVED MEMORY LOCATIONS

| Location | Contents |
|---|---|
| DIS_BUF | "D0" |
| DIS_BUF + 1 | "D1" |
| DIS_BUF + 2 | "A0" |
| . +3 | "A1" |
| . +4 | "A2" |
| DIS_BUF + 5 | "A3" |
| KEYCODE | 0 → 16H |
| KEY_STB | 0 OR 1 |
| SEL_CODE | 1011 1111B 1111 1110B |
| INT_CNT | 6 → 0 |
| ARM_KYBRD | 0 or 1 |
| DEBOUNCE_CNT | 3 → 0 |

**Table A.2**

| Key | Port 1 | Port 2 | Code |
|---|---|---|---|
| 0 | 1101 1111 | 1110 0000 | 00H |
| 1 | 1110 1111 | 1110 0000 | 01H |
| 2 | 1111 0111 | 1110 0000 | 02H |
| 3 | 1111 1011 | 1110 0000 | 03H |
| 4 | 1101 1111 | 1101 0000 | 04H |
| 5 | 1110 1111 | 1101 0000 | 05H |
| 6 | 1111 0111 | 1101 0000 | 06H |
| 7 | 1111 1011 | 1101 0000 | 07H |
| 8 | 1101 1111 | 1011 0000 | 08H |
| 9 | 1110 1111 | 1011 0000 | 09H |
| A | 1111 0111 | 1011 0000 | 0AH |
| B | 1111 1011 | 1011 0000 | 0BH |
| C | 1101 1111 | 0111 0000 | 0CH |
| D | 1110 1111 | 0111 0000 | 0DH |
| E | 1111 0111 | 0111 0000 | 0EH |
| F | 1111 1011 | 0111 0000 | 0FH |
| F0 | 1111 1101 | 1110 0000 | 10H |
| F1 | 1111 1101 | 1101 0000 | 11H |
| F2 | 1111 1101 | 1011 0000 | 12H |
| F3 | 1111 1101 | 0111 0000 | 13H |
| F4 | 1111 1110 | 1101 0000 | 14H |
| F5 | 1111 1110 | 1011 0000 | 15H |
| F6 | 1111 1110 | 0111 0000 | 16H |

## A.2.1 System Requirements

A. Accept seven-segment codes from the monitor program and pass them to the seven-segment displays.

B. Scan, debounce, and decode the keypad and pass keycodes and a keystrobe to the monitor program.

C. Ensure that a single keystroke causes a single action (so that the monitor does not accept an "old" keystroke).

D. The displays and keypad must be accessible at all times, including when a user program is running.

E. The formalism used for this design must illustrate the software design cycle and provide a model for the student's design of the monitor program.

## A.2.2 System Specifications

I. Input/Output

A. Six reserved memory locations will accept seven-segment codes from the monitor program. This program will time division multiplex the six codes across the six

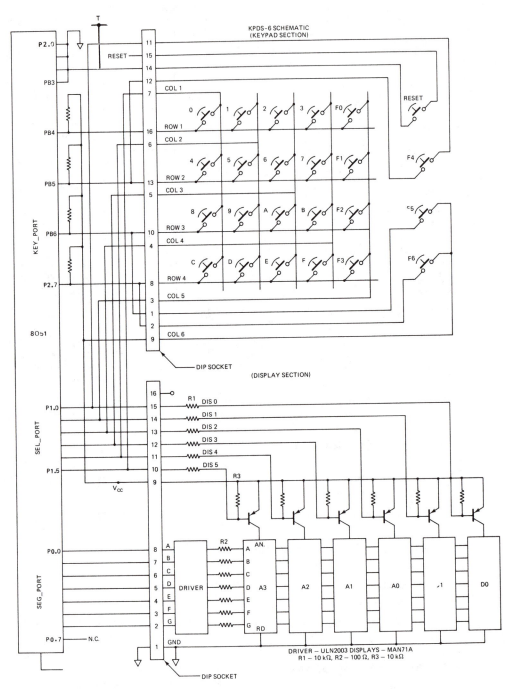

**Figure A.1**

seven-segment displays of the devry KPDS-6. Each of the displays should be on for about two milliseconds.

B. The program will check for any key closure once for each complete multiplexing of the displays (once every twelve milliseconds). If any key is closed for three consecutive scans, that keystroke is considered to be debounced, and the keypad is scanned to detect which key is depressed. The keypad is then disarmed until that key opens, so that the same keystroke is not captured twice. The program then decodes the keystroke and passes the keycode and a keystrobe to the monitor program. (The monitor program must reset the keystrobe upon acceptance of the keycode.)

C. Refer to existing hardware design for I/O port specifications.

II. Functions

A. This software will be interrupt driven through the INT interrupt request input to ensure continuous accessibility.

B. The displays will be selected one at a time by the contents of the 8051 output port P1. The corresponding seven-segment code will be passed through port P0. Each subsequent run of the program will multiplex to the next display. Reserved memory locations will keep track of the current display selected and when to re-initialize back to first display.

C. Once each complete multiplexing of the display status of the keypad is checked. This is accomplished by driving all columns of the keypad low through port P1, and checking the condition of the rows on port P2.

D. The keypad will be armed upon detection of all keys open. It will be disarmed upon successful debouncing of a keystroke. The status of keypad arming will be in a reserved memory location.

E. If any key is detected closed for three consecutive multiplexing cycles, with the keypad armed, then the keystroke is considered to be debounced. A debounce counter keeps track of number of cycles (debounce time is 24 mSEC).

F. For a debounced keystroke, the keyboard is disarmed, and the keypad is scanned. The debounce count is reinitiated (re-inited).

G. The information about a captured keystroke is decoded to an absolute (hexadecimal) keycode. The keycode is passed to the monitor program in a reserved memory location. A keystrobe is passed to the monitor program in another reserved memory location. After the monitor gathers the keycode it must reset the keystrobe.

A list of reserved memory location follows.

**DIS_BUF** THROUGH **DIS_BUF+5**
        SIX BYTES FOR PASSING DISPLAY CODES

**KEYCODE**
        PASSES KEYCODE TO MONITOR

**KEY_STB**
        PASSES KEYSTROBE TO MONITOR

**SEL_CODE**
        POINTS TO CURRENT DISPLAY

**INT_CNT**
        COUNTS DISPLAYS FOR STARTING NEW CYCLE

**ARM_KYBRD**
        GIVES STATUS OF KEYPAD ARMING

**DEBOUNCE_CNT**
        COUNTS KEY CLOSURE DETECTIONS FOR DEBOUNCE

## A.2.3 Software Modularization

```
****************************
*    INTERRUPT EXECUTIVE    *
****************************
*           TRAP            *
****************************


******************    ******************    ******************
*  KEY DECODING  *    * INITIALIZATION *    *     ERROR      *
******************    ******************    ******************
*    DECODER     *    *      INIT      *    *                *
*    LOOK_UP     *    *    STORAGE     *    *     ERROR      *
*    KEYCODES    *    *                *    *                *
******************    ******************    ******************


  ********************           ********************
  *     OUTPUT       *           *     INPUT        *
  ********************           ********************
  *    DISPLAY       *           *    CHK_KEY       *
  *                  *           *    SCAN_KYBRD    *
  ********************           ********************
```

## A.2.4 Pseudo-Code

All the following modules, procedures and routines were written by G. Thomas Huetter.

```
MODULE INTERRUPT EXECUTIVE

    PROCEDURES:
        TRAP( ; )
    EXTERNAL PROCEDURES CALLED:
        CHK_KEY( ; ROW_STATUS)
        SCAN_KYBRD( ; ROW_STATUS,COL_STATUS)
        DECODER(ROW_STATUS,COL_STATUS ; KEYCODE)
        DISPLAY( ; )

PROCEDURE TRAP( ; )

    CALLED BY:
        THIS PROCEDURE IS ACCESSED BY EXTERNAL HARDWARE THROUGH THE TRAP
        INPUT.  THE PROCEDURE IS INVOKED EVERY TWO MILLISECONDS

    PROCEDURES CALLED:
        CHK_KEY( ; ROW_STATUS)
        SCAN_KBRD( ; ROW_STATUS,COL_STATUS)
        DECODER(ROW_STATUS,COL_STATUS ; KEYCODE)
        DISPLAY( ; )
    DATA STRUCTURE:
        INPUT PARAMETERS:
            NONE
        OUTPUT PARAMETERS:
            NONE
        PROGRAM VARIABLES
            INT_CNT           (1 BYTE)
            ARM_KYBRD         (1 BYTE)
            DEBOUNCE_CNT      (1 BYTE)

BEGIN PROCEDURE

    SAVE ENVIRONMENT
    IF THIS IS FIRST DISPLAY
        THEN CALL CHK_KEY( ; ROW_STATUS)
        ELSE DO
                CALL DISPLAY( ; )
                RESTORE OLD INTERRUPT STATUS
                RESTORE ENVIRONMENT
                RETURN
            END
    IF NO KEY IS DEPRESSED
        THEN DO
                ARM KEYBOARD
                CALL DISPLAY( ; )
                RESTORE OLD INTERRUPT STATUS
                RESTORE ENVIRONMENT
                RETURN
            END
        ELSE IF KEYBOARD IS ARMED
                    THEN DECREMENT DEBOUNCE COUNTER
                    ELSE DO
                            CALL DISPLAY( ; )
                            RESTORE OLD INTERRUPT STATUS
                            RESTORE ENVIRONMENT
                            RETURN
                        END
```

```
            IF DEBOUNCE COUNTER IS NOT ZERO
                THEN DO
                        CALL DISPLAY( ; )
                        RESTORE OLD INTERRUPT STATUS
                        RESTORE ENVIRONMENT
                        RETURN
                    END
            SET DEBOUNCE COUNTER
            CALL SCAN_KYBRD( ; ROW_STATUS,COL_STATUS)
            CALL DECODER(ROW_STATUS,COL_STATUS ; KEYCODE)
            PASS CODE AND KEYSTROBE
            DISARM KEYBOARD
            CALL DISPLAY( ; )
            RESTORE OLD INTERRUPT STATUS
            RESTORE ENVIRONMENT
            RETURN

    END PROCEDURE TRAP

    END MODULE INTERRUPT EXECUTIVE




MODULE OUTPUT

        PROCEDURES:
            DISPLAY( ; )

    PROCEDURE DISPLAY( ; )

        THIS PROCEDURE TRANSFERS A BYTE FROM THE DISPLAY BUFFER TO THE
        DISPLAY SEGMENT PORT AND MUXES OVER TO THE NEXT DISPLAY.  THIS
        ALSO SELECTS THE NEXT COLUMN OF THE KEYPAD.

        CALLED BY:
            TRAP( ; )
        PROCEDURES CALLED:
            NONE
        DATA STRUCTURE:
            INPUT PARAMETERS:
                NONE
            OUTPUT PARAMETERS:
                NONE
            PROGRAM VARIABLES
                DIS_BUF                ( 6 BYTES )
                INT_CNT                ( 1 BYTE )
                SEL_CODE               ( 1 BYTE )

    BEGIN PROCEDURE

        UPDATE INTERRUPT COUNTER
        POINT TO CURRENT DISPLAY DATA
        OUTPUT SEVEN SEGMENT CODE
        UPDATE SELECT CODE
        SELECT CURRENT DISPLAY
        IF THIS IS NOT LAST DISPLAY
            THEN RETURN
        REINITIALIZE INTERRUPT COUNTER AND SELECT CODE
        RETURN

    END PROCEDURE DISPLAY

    END MODULE OUTPUT
```

```
MODULE INPUT

    PROCEDURES:
        CHK_KEY( ; ROW_STATUS)
        SCAN_KYBRD( ; ROW_STATUS,COL_STATUS)

PROCEDURE CHK_KEY( ; ROW_STATUS)

    THIS PROCEDURE CHECKS FOR ANY KEY CLOSURE. IF KEY IS CLOSED,
    ROW STATUS  (ROW_CODE) WILL CONTAIN THAT INFORMATION.

    CALLED BY:
        TRAP( ; )
    DATA STRUCTURE
        OUTPUT PARAMETERS:
            ROW_STATUS                  ( 1 BYTE )

BEGIN PROCEDURE

    DRIVE ALL COLUMNS LOW
    READ ROW STATUS
    RETURN

END PROCEDURE  CHK_KEY


PROCEDURE SCAN_KYBRD( ; ROW_STATUS,COL_STATUS)

    THIS PROCEDURE RETURNS THE "RAW" KEYCODE, WHICH IS ENCODED
    IN TERMS OF WHICH ROW AND COLUMN CONTAINS THE DEPRESSED KEY.
    CALLED BY:
        TRAP( ; )
    DATA STRUCTURE:
        OUTPUT PARAMETERS:
            ROW_STATUS              ( 1 BYTE )
            COL_STATUS              ( 1 BYTE )

BEGIN PROCEDURE
    INITIALIZE COLUMN STATUS
    DO FOREVER
        POINT TO NEXT COLUMN
        DRIVE A COLUMN LOW
        IF THIS COLUMN IS PAST LAST COLUMN
            THEN DO
                    FIX STACK POINTER
                    DISARM KEYBOARD
                    GO TO ERROR
        SAVE COLUMN STATUS
        READ STATUS OF ROWS
        IF A KEY IS DEPRESSED
            THEN RETURN
            ELSE RESTORE COLUMN STATUS
    END

END PROCEDURE SCAN_KYBRD

END MODULE INPUT


MODULE  KEY DECODING

    PROCEDURES:
        DECODER(ROW_STATUS,COL_STATUS ; KEYCODE)
        LOOK_UP(RAWCODE ; KEYCODE)
        KEYCODES( ;)
```

```
PROCEDURE DECODER(ROW_STATUS,COL_STATUS ; KEYCODE)

    THIS PROCEDURE ACCEPTS THE "RAW" KEYCODE IN TERMS OF TWO BYTES
    THAT DEFINE THE ROW AND COLUMN OF THE KEY AND COMPRESSES THAT
    INFORMATION INTO A SINGLE BYTE INTERMEDIATE (HALF_COOKED) KEYCODE,
    CALLED THE RAWCODE.  THE RAWCODE IS PASSED TO PROCEDURE LOOK_UP.

    CALLED BY:
        TRAP( ; )
    DATA STRUCTURE:
        INPUT PARAMETERS:
            ROW_STATUS              ( 1 BYTE )
            COL_STATUS              ( 1 BYTE )
        OUTPUT PARAMETERS:
            KEYCODE                 ( 1 BYTE )

BEGIN PROCEDURE
    SAVE ROW STATUS
    IF KEY IS DATA KEY
        THEN DO
                MASK COL STATUS FOR D2 THRU D5
                SHIFT RIGHT TWICE
                OR WITH ROW STATUS TO GENERATE HALF-COOKED KEYCODE
                POINT TO DATA TABLE
                INIT COUNT TO 10H
            END
        ELSE DO
                MASK COL STATUS FOR D0 THRU D3
                OR WITH ROW STATUS (HALF-COOKED KEYCODE)
                POINT TO FUNCTION TABLE
                INIT COUNT TO 07H
            END
    CALL LOOK_UP(RAWCODE ; KEYCODE)
    RETURN
END PROCEDURE


PROCEDURE LOOK_UP(RAWCODE ; KEYCODE)

    THIS PROCEDURE ACCEPTS INTERMEDIATE KEYCODE (RAWCODE), AND
    GENERATES THE FINAL KEYCODE BY COMPARISON TO KEYCODE TABLE.

    CALLED BY:
        DECODER(ROW_STATUS,COL_STATUS ; KEYCODE)
    DATA STRUCTURE:
        INPUT PARAMETERS
            RAWCODE
        OUTPUT PARAMETERS:
            KEYCODE

BEGIN PROCEDURE

    SAVE COUNT
    DO FOREVER
        COMPARE TABLE ENTRY TO HALF-COOKED KEYCODE
        IF KEYCODE AGREES
            THEN DO
                    TEST FOR FUNCTION OR DATA KEY
                    MOVE KEYCODE TO ACCUMULATOR
                    RETURN
                END
        POINT TO NEXT TABLE ENTRY
        IF PAST END OF TABLE (NO VALID KEYCODE FOUND)
            THEN DO
                FIX STACK POINTER
```

```
                      DISARM KEYBOARD
                      GO TO ERROR
         END

END PROCEDURE

PROCEDURE KEYCODES

         THE PURPOSE OF THIS PROCEDURE IS TO PROVIDE A LOOK-UP TABLE
         FOR THE PROGRAM LOOK_UP.  THE ENTRIES ARE THE HALF_COOKED
         CODES GENERATED BY PROGRAM SCAN_KYBR.

         THE CODES ARE ENTERED BY USE OF THE "HEX" ASSEMBLER DIRECTIVE.

END PROCEDURE

END MODULE KEY DECODING


MODULE INITIALIZATION

         PROCEDURES:
             INIT( ; )
             STORAGE( ; )

PROCEDURE INIT( ; )

         THIS PROCEDURE IS INVOKED UPON COLD RESTART (POWER-UP OR
         RESET BUTTON).  IT SETS UP 8155 FOR CONTROL OF KEYPAD/DISPLAY,
         INTIALIZES RESERVED MEMORY LOCATIONS, AND CLEARS 8085 REGISTERS.

         CALLED BY:
                 THIS PROGRAM IS INVOKED BY POWER-UP OR RESET BUTTON

BEGIN PROCEDURE

         INITIALIZE 8155 PORTS AND TIMER
         INITIALIZE  RESERVED MEMORY LOCATIONS
         CLEAR 8085 INTERNAL REGISTERS
         GO TO STUDENT'S MONITOR PROGRAM

END PROCEDURE

PROCEDURE STORAGE

         THIS PROCEDURE RESERVES MEMORY LOCATIONS FOR:
             A.   DISPLAY BUFFER (SIX LOCATIONS)
             B.   KEYCODE AND KEY_STB
             C.   PROGRAM VARIABLES FOR KEYPAD/DISPLAY DRIVER
                  (SEE LIST ON PAGE SIX.)

         THE LOCATIONS ARE RESERVED BY USE OF THE "DS" ASSEMBLER
         DIRECTIVE.  IF THE STUDENT WISHES TO RESERVE MORE LOCATIONS,
         THIS PROCEDURE COULD BE MODIFIED (BY ADDING MORE LOCATIONS).
END PROCEDURE

END MODULE INITIALIZATION


MODULE ERROR

         PROCEDURES:  ERROR( ; )

PROCEDURE ERROR( ; )
```

```
THIS PROCEDURE IS A TEMPORARY ERROR PROGRAM TO ALLOW THE
LINK OF THE TEST MONITOR TO PROCEED.  IT WILL CAUSE ALL
DISPLAYS TO SHOW "8'S" IF ILLEGAL KEYCODES ARE ENCOUNTERED.

    CALLED BY:
        LOOK_UP(RAWCODE ; KEYCODE)

BEGIN PROCEDURE

    DO FOREVER
        DISPLAY ALL "8'S"
    END

END PROCEDURE

END MODULE ERROR
```

### A.2.5 Desigh Walkthrough

Each design language procedure would be checked for logical errors by ''walking through'' the steps, that is executing the procedure in a thought experiment. The design walkthrough would precede the assembly language implementation.

The complete design language could be implemented prior to writing any of the assembly language procedures, or alternatively, assembly language procedures could be done as soon as the corresponding design language procedure is written and walked through.

### A.2.6 Procedure Calling Tree

In this diagram the parameters are denoted as input parameters above the lines and output parameters below the lines.

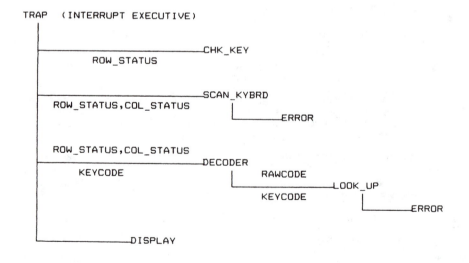

The procedure INIT is invoked upon power-up or cold restart (reset). The "procedures" STORAGE and KEYCODES are not executable code, rather they are data set-up procedures.

### A.2.7 Assembly Language

An assembly language implementation is left up to the student. The order of the assembly language procedures is the same as the order of the design language procedures. The procedures can be linked in any order, so long as the procedure trap is first in the link list and is located at address zero.

### A.2.8 Verification

The assembly language procedures can be executed on a simulator or an emulator and checked for proper operation.

# Introduction to Assembly Languages

## B.1 INTRODUCTION

A computer requires a step-by-step list of instructions called a program that tells it exactly what to do. However, the only thing a computer can actually read and process is a binary number. A microprocessor (or any computer) is designed to recognize a group of binary numbers as its *native* or *primitive instruction set*. Such an instruction set makes up the *machine language* for a processor. Although pure numbers are fine for machines, people are more comfortable with words and names. As we will discuss, an assembly language is a machine language with words replacing the numbers.

### B.1.1 Mnemonics

In the very early days, programs were actually entered into a computer as binary numbers using toggle switches on the machine's front panel, or the computer might read the numbers as holes punched into cards or paper tapes. Either way, such programming could not be called user-friendly. To write programs that could be read and understood by people, programmers used short words or letter combinations to represent the operation codes *(op-codes)*. For example, the word ADD might be written to stand for the binary number 11000110, which would be the machine code for the addition instruction. Such short words are called *mnemonics,* from a Greek word meaning "an aid to memory."

### B.1.2 The Assembler: A Translator

At first, programmers had to translate the mnemonics back into binary numbers by hand and then enter them into the computer. It wasn't long, however, before computers were enabled to do the tedious job of translation. Using a device with an alphanumeric keyboard, such as a teletype, a person could enter programs written with mnemonics directly into the computer. Pressing a key on a teletype machine produces a binary code representing the character on the key. Thus, a mnemonic such as ADD would be represented by three binary numbers. A program written with words is said to be in *symbolic* form.

After the program was entered in symbolic form, a special translator program was run that would recognize the binary numbers making up the mnemonics and produce the binary op-codes. The translator program is called an *assembler,* and programs written in the mnemonic form are called *assembly language* (or *assembler language*) programs.

### B.1.3 Assembly: A Low-Level Language

An assembly language has a one-to-one correspondence to the underlying machine language. The main difference is that the assembly language allows the numeric op-codes to be replaced by mnemonics and the numeric operands to be replaced by symbolic names. The manufacturer of the microprocessor also defines (and copyrights) the mnemonics for its products. Each microprocessor has its own unique assembly language, although some (e.g., the 8085 and the Z80) may, to a certain extent, be compatible at the machine code level. Assembly languages are classified as *low-level* languages, in contrast to *high-level* languages such as BASIC or PASCAL.

### B.1.4 Why Use Assembly Language?

Programs written in a high-level language can be run on computers built with different microprocessors, whereas programs written in assembly language can be executed only on computers built with the microprocessor for which the assembler was designed. Also, high-level languages contain instructions that allow a programmer to do, in a few lines of code, operations that would take many lines of code in assembly language. So why use assembly? There are several reasons:

1. Assembly language gives the user full access to the power of the processor. Because high-level languages must run on any machine, they do not allow the programmer access to processor-specific features. A high-level language defines a *virtual machine,* which hides features of the real processor from the programmer.

2. Assembly language allows direct control of hardware features such as I/O ports, CPU registers, and memory. High-level languages usually do not allow such control. Much of what is called *systems programming* is done in assembly.

3. Assembly language allows a skilled programmer to write "tight code" — that is, a program written with the fewest machine code instructions and able to execute in the least amount of time. Programs written in a high-level language must still be translated down to machine code by a translator program called a *compiler.* However, even the best compilers are not as efficient as a skilled human programmer (at least not yet). For applications requiring very compact code, such as those using a single-chip controller with on-board ROM, assembly is often preferred.

4. Assembly language allows a programmer to fine-tune the timing of a program. Interrupt-driven real-time applications often require precise control over the execution

times of critical parts of the program, which is often difficult to achieve in a high-level language.

## B.1.5 Editors and Source Files

Before writing a line of code, a good programmer will have already designed the program using flowcharts, pseudo-code, or some other technique. (Program design is discussed in Chapter 5.) After designing the program, the programmer writes the actual assembly language code ("codes it up") using a program called an *editor*. An editor allows the programmer to enter a program from a keyboard while monitoring it on a CRT display; he or she can then make any additions, deletions, or corrections and save the program on a hard or floppy disk. The stored program can be retrieved by the editor for further changes.

Editors in common use with microprocessors use the ASCII code to represent the letters, numbers, figures, and punctuation marks on the keyboard. A *programmer's editor* uses the ASCII code exclusively. Some word processor types of editors mix special non-ASCII codes with the standard ASCII characters. Such word processor editors should not be used to write programs because microprocessor assemblers will only accept ASCII character codes. The file that the editor stores on disk is called a *source file*.

## B.1.6 Object Files, List Files, Loaders, Linkers

The job of the assembler is to read the source file and create a new file containing a translation of the source file. The assembler does not alter the source file. The new file, often called an *object file*, created by the assembler contains the numeric codes required by the microprocessor. In addition, the assembler will generate a *list file*, which can be printed out. A list file contains each line of source code together with the machine code translation.

If the object file can be loaded into the microprocessor's memory and run "as is," then it is called an *absolute file* or a *binary file*. Often it is the case that the numeric addresses of memory references in a program depend on where in memory the program will be placed for execution. However, the numeric value of the beginning address may not be known at the time the program is written but will be determined when the program is actually loaded into memory. In such a case the object file must be created in a form called a *relocatable file*, and a program called a *loader* will adjust all program addresses to their absolute numeric value as it loads the program into memory. After a relocatable object file has had its address references adjusted and is ready for execution, it is called a *load module*.

Large programs must be broken down into smaller parts (or modules) so that a single programmer can work on manageable pieces of code or so that a team of programmers can work simultaneously. The result is a collection of relocatable object files that must be combined into one final program. The combining is done by a program called a *linker*, which makes sure all references to memory locations contain the correct addresses. Sometimes the linker and the loader are combined into one program. Also, collections of useful routines that can be used in many different programs are brought together into *libraries*, from which they can be linked into new programs as the need arises.

Because large programs often consist of many modules, changes made to one module may require other modules to be changed as well. A *configuration control* program can be used to keep track of all such changes automatically. In addition, it can reassemble and relink the modules into a new version of the program and it saves previous versions in an *archive*.

### B.1.7 Development Systems

The computer used to write and assemble a program is part of a *development system* or a *work station*. A development system provides the hardware and software tools needed to write and debug programs. Sometimes the development system is used to write software for a different processor than the one contained in the development system itself. In such a case, a *cross-assembler* is required. A cross-assembler will run on the development system processor but will produce as its output machine code to run on a different processor. The microprocessor-based piece of equipment that will eventually run the program produced by the cross-assembler is called the *target system*. The 8085 assembler available with this book is a cross-assembler. It runs on an IBM personal computer or clone. The single-board computer described in Appendix A (or any similar board) can be used as a target system.

A development system usually has a piece of equipment for transferring or *downloading* the output of a cross-assembler into the memory of the target system. Such equipment may be capable of writing into the on-board memory of a single-chip computer as well as into an EPROM device. The equipment that programs the microprocessor memory may require an absolute load module or, alternatively, a *hex file*.

Many development systems include *emulation* as a powerful design tool. One version, called *in-circuit emulation*, allows the development system to take the place of the microprocessor in the target system. A cable plugs into the socket where the microprocessor would go and connects the development system to the target system. During program execution, special *debugger* software allows the programmer to examine and alter the contents of CPU registers, flags, and memory locations. Execution can be single-stepped, and modifications to the program can be made and tested instantly.

A *simulator* is a program that models in software the behavior of hardware, allowing the user to find errors in a design before any hardware is built. However, because real hardware can sometimes behave in unexpected ways, the usefulness of a simulation is limited by how faithfully it represents the hardware.

### B.1.8 Assembler Passes and Errors

The typical assembler is called a *two-pass* assembler, meaning that it reads the source file twice. Two passes are required because some information needed to complete the translation of the beginning of the program may not be found until the end of the program. Information is gathered during the first pass, and the actual translation is done during the second pass. Figures B.1 and B.2 show the steps involved in each pass.

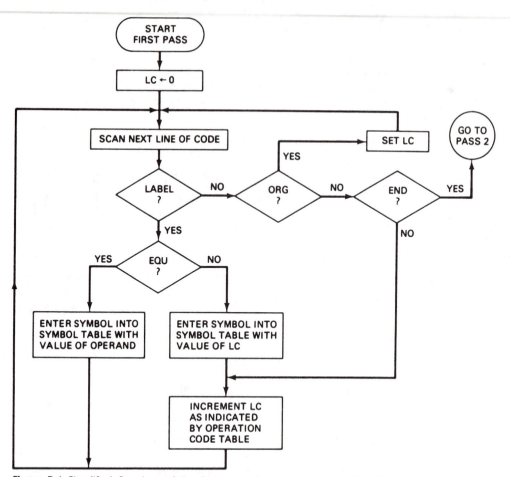

**Figure B.1** Simplified flowchart of the first pass of a two-pass assembler. (*Source:* Kenneth Short, *Microprocessors and Programmed Logic,* © 1981, p. 162. Reprinted by permission of Prentice-Hall, Inc., Englewood Cliffs, N.J.)

A mistake in using the language properly is called a *syntax* error. An example of a syntax error is misspelling a mnemonic. The usual practice is for the assembler to report all errors in the listing file. The generation of a usable object file is voided by errors. *Semantic* errors, or errors in the design of the program, will not be caught by the assembler but must be found by testing.

### B.1.10 The Programming Model

The term *programming model* refers to the way a programmer must view a computer. Whereas a hardware designer must be concerned with such details as bus loading and power supply voltages, the programmer is concerned with those features of the processor that relate to writing code. Besides the instruction set, the programming model includes

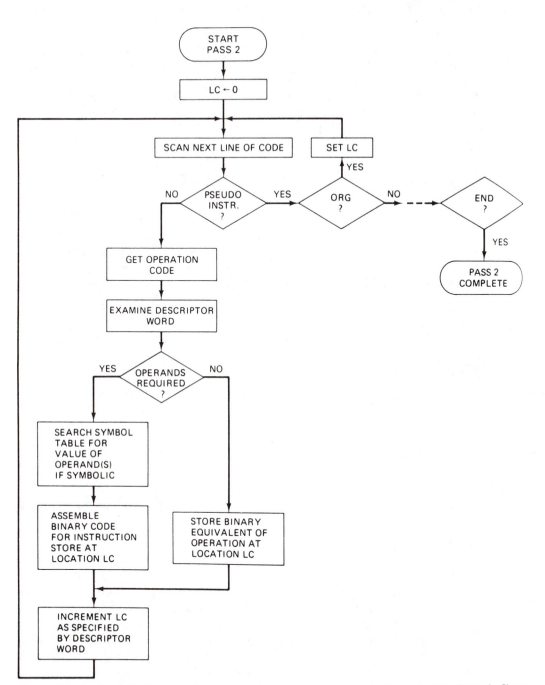

**Figure B.2** Simplified flowchart of the second pass of a two-pass assembler. (*Source:* Kenneth Short, *Microprocessors and Programmed Logic*, (© 1981, p. 163. Reprinted by permission of Prentice-Hall, Inc., Englewood Cliffs, N.J.)

such things as the sizes and names of all the registers, the memory map showing the address ranges of RAM and ROM, I/O port addresses, handshaking sequences, and details of any special processor features that may be used in a program.

Assembly language gives the programmer full access to all features of the processor, so it is essential that the programmer fully understand what the processor can and cannot do. The programming model is the image of the machine that the programmer must keep in mind while designing and writing the program.

## B.2 INSTRUCTION STATEMENTS

A typical *statement* in an assembly language consists of an optional *label,* followed by an operation mnemonic, followed by one or more operands. A comment may follow on the same line, separated from the statement by a semicolon. Comments are optional, but it is good programming practice to include enough comments to make the program as self-documented as possible for the benefit of whoever has to modify the program during its lifetime. Even programs you write yourself can seem unintelligible to you six months later if they are not well documented. Typically, much more time is spent on *maintenance* (i.e., modifying and debugging) than on writing programs in the first place.

### EXAMPLE B.1

Some example statements:

```
;    THIS IS A COMMENT
MAIN:   MVI A,45;   LOAD ACCUMULATOR
LP1:    INR A
        JMP LP1
```

MAIN: and LP1: are labels; MVI, INR, and JMP are mnemonic op-codes; A, 45, and LP1 are operands.

### B.2.1 Labels and Names

Note that labels must begin with a letter and terminate with a colon. They can be used as operands in other statements because a label is actually a symbolic representation of a numeric memory address. The assembler will determine the numeric value of a label during pass 1 and store it in the assembler's *symbol table*. During pass 2, when the assembler encounters the label again, it will look up the value in the symbol table as it translates the symbolic assembly language into numeric machine language.

A *name* is similar to a label in that it is a symbolic representation of a numeric value. Names are associated with a value in an *equate statement*. An equate is not a part of the microprocessor's instruction set but is a feature of the assembler itself. Such features, called pseudo-operations (or assembler directives), are discussed in Sec. B.4.

**EXAMPLE B.2 EQU**

An example equate:

```
FIVE   EQU   5
```

Unlike labels, names do not usually end with a colon. After they have been defined, names can be used as operands in instructions:

```
MVI   A,FIVE
```

Both names and labels must begin with a letter and usually may not contain spaces, punctuation marks, or special characters such as # or @. The assembler user's manual will also specify the maximum number of characters that names and labels may contain, as well as the maximum number of names and labels that the symbol table will hold. Another limitation on names and labels is that they cannot be the same as any key words. *Key words* are typically the names of flags and registers, op-code mnemonics, and pseudo-operations.

## B.2.2 Operands

Operands can be numeric or symbolic. Symbolic operands may be labels or names, or they may be fixed symbols defined by the assembly language itself, such as the designations of registers in the CPU. See Example B.3

**EXAMPLE B.3 OPERANDS**

```
MOV   A,B
MVI   A,FIVE
```

The first instruction moves the content of the B register into the A register; the assembler will know what A and B refer to. The second instruction moves the value of the symbol FIVE into the A register. The assembler will recognize A, but will only know the value of FIVE if the name has been defined in an EQU statement.

## B.2.3 Numbers and Radix

Most assemblers allow numbers to be represented in several bases, with decimal usually the normal or *default* base. The term *radix* is often used for the base of a number; a decimal number has a radix of ten (base 10). Other common bases are hexadecimal (base 16), octal (base 8), and binary (base 2).

The assembler determines the base of a number by looking for a letter at the end of the number. No letter indicates the default base, D indicates a decimal number, H indicates a hexadecimal number, O or Q indicates an octal number, and B indicates a binary number. Thus, the number 10 is ten, 10H is sixteen, 10Q is eight, and 10B is two.

Care must be used with hexadecimal (hex) numbers because of the letters A through F. To convince the assembler that something is a number, it must start with a digit in the range 0 to 9.

## EXAMPLE B.4 WRITING HEX NUMBERS

The hex number F2H looks like a name to the assembler, and it will try to find the value in the symbol table. Therefore, F2H should be written as 0F2H to avoid an error.

Also, if you forget the H at the end of a hex number, the assembler may or may not see it as an error.

## EXAMPLE B.5 WRITING BINARY NUMBERS

Assuming the default base is decimal, the number 1A will be detected as an error, but the hex number 1B will be read as if it were a valid binary number.

### B.2.4 Characters and Strings

In a computer, all symbols are ultimately represented by numbers, including all the numbers, letters, and punctuation marks on a keyboard. The numbers used to represent keyboard characters are *codes*. The code most commonly used with microcomputers is *ASCII* (American Standard Code for Information Interchange). Another common code is *EBCDIC* (Extended Binary Coded Decimal Interchange Code), used by IBM in its larger machines.

To represent a character code in assembly language, enclose the character between single quote marks (apostrophes). See Example B.6.

## EXAMPLE B.6 CHARACTERS

An assembler will read 5 as the number 05H but will read '5' as 35H, which is the numeric value of the ASCII code for the character 5. In the same manner, 'A' will evaluate to 41H, which is the ASCII code for the letter A.

A *string* is a group of characters enclosed between single quotation marks such as 'this is a string'. Note that when counting characters in a string, the spaces (blanks) must also be counted. See Example B.7.

## EXAMPLE B.7 STRINGS

The string 'this is a string' contains 16 characters.

### B.2.5 Expressions

It is often useful to let the assembler calculate the values for symbols using mathematical expressions. An *expression* is made up of arithmetic combinations of numbers and symbols such as $2+3*VAR$, where VAR is a previously defined name. The normal operator precedence is usually observed, meaning multiplication and division are done before addition and subtraction. Terms inside parentheses will be done first.

## EXAMPLE B.8 EXPRESSIONS

The expression 3*1+2 will evaluate to 5; 3*(1+2) will evaluate to 9.

Expressions can be used as operands in statements and in pseudo-operations. See Example B.9.

## EXAMPLE B.9 EXPRESSIONS AS OPERANDS

```
MVI    A,3*FIVE-1
TWO    EQU 2
FIVE   EQU 3+TWO
MVI    B, 'A'+1
```

Note that TWO had to be defined before it could be used in an EQU statement, but FIVE was used in an instruction before it was defined. The reason is that EQU statements are evaluated during pass 1 in the order they appear, whereas operands in instructions are evaluated during pass 2. Note also that the last statement in the example will put the value 42H into the B register when it is executed.

# B.3 ADDRESSING MODES

An important consideration in judging the power of a processor is the number of different ways instructions can access the memory, that is, the processor's *addressing modes*. In this section we examine several typical methods. The term *load* is used to mean that the processor is reading data from memory into a register; the term *store,* writing data into memory.

## B.3.1 Register Addressing

In processors with multiple general-purpose registers, the fastest way to move data around is from register to register. *Register addressing* uses the predefined names of the registers in the instruction.

## EXAMPLE B.10

The 8085 instruction MOV A,B moves a copy of the contents of register B into register A. The operand names A and B are part of the assembly language.

## B.3.2 Direct Addressing

In *direct addressing* the instruction contains the complete address of the memory location it is trying to reference. The address may be in numeric or symbolic form. Instructions using direct addressing are limited to the one memory location they specify, so loading from a hundred sequential locations would require a hundred different instructions.

### EXAMPLE B.11

The two 8085 instructions LDA LABL1 and LDA 1234H are equivalent if LABL1 appears in the program as the label of the location with address 1234H or has been defined as 1234H with an EQU.

## B.3.3 Immediate Mode

*Immediate mode* is a method of putting a constant value into a register or memory location during program execution. It is a modification of addressing methods such as register or direct. For example, the 8088 instruction to load the number 12H into register AL is MOV AL,12H. In some assembly languages (e.g., 8085) the form of the mnemonic is changed to indicate immediate mode, so the equivalent move instruction would be written MVI A, 12H. Once the program is assembled, immediate operands have a fixed value; they *cannot* be changed by the program.

## B.3.4 Indirect Addressing

*Indirect addressing* gets around the limitation of direct addressing by using a pointer to hold an address. The pointer may be a memory address or a register. Instructions can then refer to that pointer instead of the actual address. When the processor executes the instruction, it will use the number in the pointer as the address. Thus, the same instruction can be used to access many different locations simply by changing the pointer contents.

### EXAMPLE B.12

In the following Z80 instructions

```
LD   HL, 1000H
LD   A, (HL)
```

the HL register pair is being used as a pointer and is initialized to hold address 1000H. The instruction LD A,(HL) will move a copy of the contents of memory location 1000H into register A. Note the parentheses around HL. The same instruction without the parentheses would mean "move the contents of the HL register pair into the accumulator," which can't be done. Thus, notation such as parentheses is needed to show when a register is to be used as a pointer. The equivalent 8085 instructions are

```
LXI   H, 1000H
MOV   A, M
```

### B.3.5 Paged Addressing

*Paged addressing* can be considered a variation on indirect addressing. Let's assume that the processor uses a 16-bit address, giving a maximum address space of 64K bytes (note that 1K is 1024). A 64K memory can be thought of as 256 *pages,* with each page consisting of 256 bytes. An 8-bit number is required to specify a page while another 8-bit number can specify a byte within a page.

In processors supporting paged addressing, an 8-bit *page register* is used to hold the page number. The contents of the page register can be altered by the program. An instruction can supply the 8-bit byte address either directly or indirectly, and the processor will concatenate the two 8-bit numbers into a 16-bit memory address when the instruction is executed.

### B.3.6 Base, Index, Displacement

*Base displacement* is a variation on indirect addressing that is somewhat similar to paged addressing but much more powerful. A *base register* (under software control) is used to hold a full-length address. Instructions that use the base register may supply a *displacement* number that will be added to the contents of the base register at execution to form a memory address. The displacement may be an immediate operand or the contents of another register.

Some processors contain an *index register,* which can be used with the displacement register. The index register also is loaded with an address. Instructions can then reference both the base and index registers as well as supply a displacement. When such an instruction is executed, the memory location address will be the sum of the base, the index, and the displacement. To increase the power of indexed addressing, some instructions may cause the processor to increment (or decrement) the contents of the index register automatically when they are executed. The 8086 uses base, index, and displacement.

### B.3.7 Relative Addressing

*Relative addressing* is similar to base-displacement addressing with the program counter taking the place of the base register. An instruction using relative addressing will supply a signed 8-bit *offset,* which is added to the contents of the program counter. In a signed number, the MSB is the sign bit and the lower 7 bits are the magnitude, so the range of addresses is within 128 bytes either forward or backward from the location of the relative instruction itself.

The most common use of relative addressing is a JUMP RELATIVE instruction. Some assemblers will allow the programmer to use a generic JUMP instruction. If the target of the jump is within relative range, the assembler will produce the op-code for a relative jump; otherwise, it will produce the code for an absolute jump. Relative jump instructions are shorter and execute faster than absolute jumps. The Z80 and 8051 use relative jumps; the 8085 does not.

### B.3.8 Implied Addressing

In *implied addressing,* the operands, which are usually registers, are fixed and are not explicitly specified in the instruction. An example is the 8085 instruction CMA, which complements the contents of the accumulator.

## B.4 PSEUDO-OPERATIONS AND DIRECTIVES

*Pseudo-operations*—sometimes also called *assembler directives*—look like assembly language instruction statements, but actually they are instructions to the assembler itself. They do not take up memory space in the final program. Psuedo-operations allow the programmer to control certain operations during the assembly process.

The mnemonic op-codes in an assembly language are determined by the designers of microprocessors and are different from processor to processor. Pseudo-ops are more or less standard for all assemblers, although any given assembler may use a variation of a particular pseudo-op or not use it at all. There are many possible assembler directives. In this section we look at some of the more important ones. Note that some assemblers precede directives with a period, such as .ORG, to differentiate them from instructions.

### B.4.1 ORG and END

ORG stands for "origin" and allows the programmer to determine the address of the first instruction in a program. If there is no ORG statement at the beginning of a program, the assembler will assume a default value, typically 0000. As each instruction is assembled, it will be assigned by the assembler to the next available address unless another ORG statement is encountered. Some assembly languages use LOC (for "location") in place of ORG.

#### EXAMPLE B.13 USE OF ORG

```
        ORG   1000H;
L1:     MOV   A,B;
        ORG   1500H;
L2:     JMP   L1;
```

The assembler will associate the label L1 with address 1000H, which is where it assumes the translated version of MOV A,B will eventually be loaded in memory. The label L2 will be associated with address 1500H. The assembler assumes nothing about all the empty locations between the two instructions shown.

The END statement tells the assembler where to stop assembling. There must be only one END directive, and it should be the last line in the program because the assembler will ignore everything after it. Omitting the END will cause an error in assembly.

## B.4.2 EQU and SET

As discussed in Sec. B.2.1, EQU allows the programmer to bind a value to a name. Some assembly languages use an equal sign (=) in place of EQU. However, once a name is defined with an EQU, it may not be redefined.

**EXAMPLE B.14 USE OF EQU**

```
VAR   EQU  2;   definition
MVI   A,VAR
VAR   EQU  3;   redefinition (error)
MVI   B,VAR
```

These lines of code would cause the assembler to report an error.

SET is similar to EQU, except that it allows a name to be redefined.

**EXAMPLE B.15 USE OF SET**

```
VAR   SET  2    temporary definition
MVI   A,VAR
VAR   SET  3    new definition
MVI   B,VAR
```

This code allows the name VAR to be used twice with different values.

## B.4.3 Memory Allocation: DS, DB, DW

Memory locations can be defined and initialized to data values by the use of *define storage* (DS), *define byte* (DB), and *define word* (DW) statements.

DS allows the programmer to set aside a block of memory without putting anything into it. See Example B.16.

**EXAMPLE B.16 USE OF DS**

```
       ORG 1000H ; anything after
STOR:  DS 5      ; a semicolon is
       MOV A,B   ; a comment
```

This code will set aside 5 bytes of memory starting at address 1000H. Also, the address of the first byte is associated with the label STOR. The instruction MOV A,B will be stored at address 1005H.

DB allows the programmer to initialize 1 or more bytes of memory to specific values. See Example B.17.

## EXAMPLE B.17 USE OF DB

```
        ORG  1000H;
TWO     EQU  2;
L1:     DB   3+TWO;
        DB   '2';
MSG:    DB   'HELLO';
```

The EQU statement does not take up memory space in the program because it is a pseudo-op. Address 1000H, associated with the label L1, will contain the numeric value 05. Address 1001H will contain the character '2', which is numeric value 32H in ASCII. Locations 1002H through 1006H will contain the values 48H ('H'), 45H ('E'), 4CH ('L'), 4CH ('L'), 4FH ('O'). The label MSG will have the value 1002H, the address of the first character in the five-character string.

The DW directive is similar to DB except that a 2-byte word is initialized instead of a single byte. Usually DW does not allow character strings, but will allow expressions. See Example B.18.

## EXAMPLE B.18 USE OF DW

```
        ORG  2000H;
ADR1:   DW   1234H;
ADR2:   DW   ADR1+5;
```

Label ADR1 is equal to 2000H and ADR2 is equal to 2002H. The content of memory starting at location 2002H is 1239H.

Some processors (e.g., the 8085) store the high-order byte in the higher memory location, whereas other processors (e.g., the 8051) do the reverse. The result is that

```
DW   1234H;
```

may or may not be equivalent to

```
DB   12H;
DB   34H;
```

depending on the processor for which the assembler was written.

It is important to note that when writing programs that will be executed from ROM, you cannot use DB or DW to initialize the RAM. You should use immediate mode instructions instead.

### B.4.4 Code and Data Segments: CSEG, DSEG

Relocatable assemblers (i.e., those that produce relocatable object files) require directives concerning address references that are not required in absolute assemblers. Relocatable assemblers allow the programmer to partition the program memory space into two regions: a *code segment* and a *data segment*. Special directives such as CSEG and DSEG are used to tell the assembler what lines of the program belong in which segment.

## EXAMPLE B.19 CSEG AND DSEG

```
        CSEG
        LXI  B,VAR1;
        DSEG
VAR1:   DB     12;
VAR2:   DB     34;
        CSEG
        JMP    NEXT;
```

In the final load module, the JMP instruction will follow the LXI instruction in memory since they are both in the code segment. Although it appears that JMP and LXI are separated by 2 bytes of defined data, they are actually sequential because the two DB statements are in the data segment, a completely different part of memory.

## EXAMPLE B.20 ADDRESSES IN SEGMENTS

```
CSEG
ORG   0100H;
DSEG
ORG   0100H;
```

It seems that the code and data segments have been assigned to the same place in memory. However, in a relocatable assembler, the ORG statements are relative to the beginning addresses of the segments, which, in turn, are determined by the loader program. The programmer can tell the loader at what addresses the segments will start. Thus, ORG 0100H in the data segment might translate to actual address 1823H in memory and ORG 0100 in the code segment might correspond to address 0400H in memory.

Another reason for separate code and data segments has to do with the physical memory. In the target system, such things as instructions and constants (ROMable code) can be placed in ROM, whereas changeable variables (data) must be kept in RAM. The use of CSEG and DSEG in the source program makes it easy to separate ROMable from non-ROMable code. As mentioned, do not use DB or DW to initialize RAM data in the DSEG if the code will be executed from ROM in the target system.

### B.4.5 PUBLIC and EXTERNAL

When relocatable object files (or modules) are linked together to form a larger program, it is often the case that a label or variable name will be defined in one module but used in several others. Such *external references* can be resolved by the linker but require the use of directives to prevent the assembler from assuming that a name not found in the symbol table is an error. The exact mnemonics used depend on the assembler, but they are usually called *PUBLIC* and *EXTERNAL*. The term *GLOBAL* is the same as PUBLIC.

A public directive usually precedes a list of symbols, as in Example B.21.

**EXAMPLE B.21**

```
PUBLIC    VAR1, VAR2, LABL1
```

The values of the symbols must be defined in the same module that declares them public. The relocatable assembler will save the public symbols in a special section of the object file so that they can be referenced by name in other modules. Any given symbol can be declared public in only one module.

For a public symbol to be used in other modules, it must be declared as external in each module that wants to use it (except for the module that declared it as public). A symbol that has been declared as public in one module can be used in a second module with a different meaning as long as the second module defines it without declaring it as either public or external.

**EXAMPLE B.22**

```
EXTERNAL    VAR1, VAR2
```

## B.5 MACRO ASSEMBLERS

Often, certain sequences of instructions are found to form a block of code that is repeated at many places in the same program. As a convenience, the programmer may wish to associate a symbolic name with the instruction sequence, and everywhere the sequence occurs it can be replaced with its name. A symbolic name that represents such a block of code is called a *macro,* and assemblers that support macros are called *macro assemblers*.

### B.5.1 Macros and Subroutines

Macros should not be confused with subroutines. A subroutine is a processor function and involves such overhead as saving register contents, pushing and popping the return address, and branching to a different part of memory to find the subroutine instructions— all of which happen when the program is executed. In contrast, macros are a feature of the assembler. When the assembler finds a macro in the program it expands it, meaning that the assembler replaces the macro name with the block of code it represents.

From a programming point of view, it may seem that much of what a subroutine does can be done by a macro. However, note that the use of subroutines makes a program shorter but slower, whereas the use of macros makes it longer but faster. A useful technique is to combine macros and subroutines. A subroutine call can be embedded in a macro together with all the instructions needed to save and restore the environment—all the registers and flags that must be preserved from alteration by the subroutine. Such a technique can save the programmer a lot of tedious coding and also prevent errors.

### EXAMPLE B.23 A SIMPLE MACRO

Assume TOTAL is a macro name. The following

```
MVI   A, VAR1;
TOTAL
MVI   A, VAR2;
```

would expand to be

```
MVI   A, VAR1;
PUSH  PSW      ; part of TOTAL
CALL  SUM      ; part of TOTAL
POP   PSW      ; part of TOTAL
MVI   A, VAR2;
```

assuming that TOTAL had been defined to mean the three instructions shown.

### B.5.2 Macro Definition

The exact form of a macro definition depends on the particular assembler being used, so what follows can be considered as generic. The assembly language user's manual will give the correct form.

A macro is defined by creating a *template* (also called a *prototype*) associated with the macro name. The *body* of the macro is the group of instructions in the template that will be substituted at every place where the macro name appears in the program. The beginning and end of the macro template are delimited by a pair of directives such as MACRO and ENDM.

An important part of a macro definition is the list of *formal parameters*. When the macro is *called* (or *invoked*) in the program, the call supplies actual parameters that will be substituted for the formal parameters when the macro is *expanded*. Look at Example B.24.

### EXAMPLE B.24 MACRO PARAMETERS

```
SUM     MACRO   $X,  $Y,  $Z
        MOV     A,   $X
        ADD     $Y
        ADD     $Z
        MOV     $X,  A
        ENDM
```

Note that the name of the macro, SUM, appears on the same line as the MACRO directive, which is, in turn, followed by the formal parameter list. Also note that the formal parameters use the $ character to mark them as special symbols (different assemblers may use other characters, such as # or %). Normally, a macro definition may not include another macro definition.

Now suppose that the macro in Example B.24 is invoked in a program with the statement

```
SUM   B, C, D;
```

It will be expanded by the assembler into

```
MOV   A, B
ADD   C
ADD   D
MOV   B, A
```

Note how the formal parameters of the template have been replaced by the actual parameters of the invocation. Different invocations can use different actual parameters.

### B.5.3 Labels in Macros

If a label appears in the body of a macro, then repeated invocations would lead to the same label being used for different addresses, a redefinition error. To get around the problem, the macro definition will include a special string in labels such as #SYM. When the macro is invoked, #SYM will be replaced in the expansion by a four-digit hexadecimal number. The number will be incremented by 1 at every call, so that no two labels will be the same.

#### EXAMPLE B.25 LABELS IN MACROS

```
DELAY   MACRO
        PUSH   A
        MVI    A, 100H
L#SYM   DEC    A
        JNZ    L#SYM
        POP    A
        ENDM
```

The first time the macro DELAY is used, L#SYM will be replaced by L0001. In a second invocation it will be replaced by L0002, in a third by L0003, and so forth. Such special strings should not appear in the list of formal parameters.

## B.6 CONDITIONAL ASSEMBLY

It is often the case that several different versions of a program are required. For example, several models of a microprocessor-based product may be sold with each model adding a few options onto a basic design. The software for each model would be basically the same, but some of the code would depend on the options. Instead of maintaining a separate program for each model, it is possible to maintain a single program with several *conditional* sections of code.

Appendix B: Introduction to Assembly Languages

### B.6.1 Conditional Directives

As the name implies, code designated for *conditional assembly* will be translated into machine code only if certain conditions are true when the program is assembled. Special directives, such as IF and ENDIF, are used to indicate the beginning and end of the conditional section of code. Conditional blocks may be *nested,* meaning that one conditional block may be completed inside another.

**EXAMPLE B.26 CONDITIONAL ASSEMBLY**

A conditional block:

```
IF    COND1
MOV   A,B
ENDIF
```

The symbol COND1 will be evaluated at assembly. If it evaluates to true, then the line MOV A,B will become part of the program. If COND1 evaluates to false, then MOV A,B will be ignored. In this context false usually means a numeric value of zero and true any nonzero value.

### B.6.2 Compound Conditions

The condition following the IF directive can be compounded from several symbols using both arithmetic expressions and logic operators. Use of an arithmetic expression is given in Example B.27.

**EXAMPLE B.27**

```
IF   2*VAR-10
MVI  A,5
ENDIF
```

If 2*VAR-10 equals zero, then the instruction will be ignored.

Use of logic operators is shown in Example B.28.

**EXAMPLE B.28**

```
IF   (M1  AND  M2)  OR  (Q3  GT  5)
MVI  A,5
ENDIF
```

For the condition to be true, both the variables M1 *and* M2 must be true at the same time *or* the variable Q3 must be greater than 5.

Common relational operators are the following:

**EQ**   equal        Example:  `IF (X EQ 5)`   is true if $X = 5$.

**GT**   greater than   Example:  `IF (X GT 5)`   is true if $X = 6$ or more.

**LT**   less than     Example:  `IF (X LT 5)`   is true if $X = 4$ or less.

**NOT**  logic inversion  Example:  `IF NOT (X GT 5)`  is true if $X = 5$ or less.

# MCS®-51 Programmer's Guide and Instruction Set

The information presented in this chapter is collected from the MCS®-51 Architectural Overview and the Hardware Description of the 8051, 8052 and 80C51 chapters of this book. The material has been selected and rearranged to form a quick and convenient reference for the programmers of the MCS-51. This guide pertains specifically to the 8051, 8052 and 80C51.

The following list should make it easier to find a subject in this chapter.

## MEMORY ORGANIZATION

## PROGRAM MEMORY

The 8051 has separate address spaces for Program Memory and Data Memory. The Program Memory can be up to 64K bytes long. The lower 4K (8K for the 8052) may reside on-chip.

Figure 1 shows a map of the 8051 program memory, and Figure 2 shows a map of the 8052 program memory.

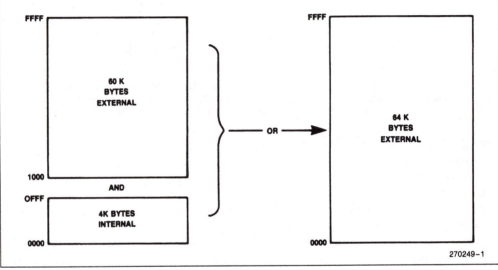

**Figure 1. The 8051 Program Memory**

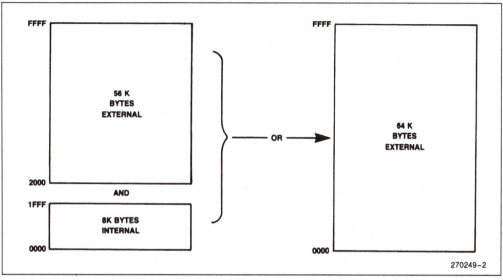

**Figure 2. The 8052 Program Memory**

## Data Memory:

The 8051 can address up to 64K bytes of Data Memory external to the chip. The "MOVX" instruction is used to access the external data memory. (Refer to the MCS-51 Instruction Set, in this chapter, for detailed description of instructions).

The 8051 has 128 bytes of on-chip RAM (256 bytes in the 8052) plus a number of Special Function Registers (SFRs). The lower 128 bytes of RAM can be accessed either by direct addressing (MOV data addr) or by indirect addressing (MOV @Ri). Figure 3 shows the 8051 and the 8052 Data Memory organization.

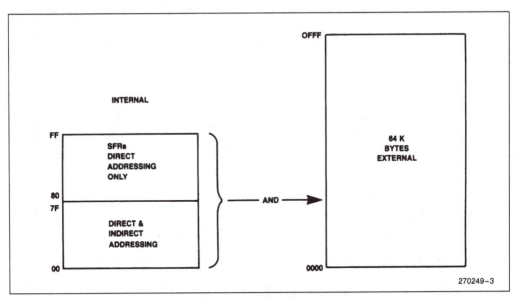

Figure 3a. The 8051 Data Memory

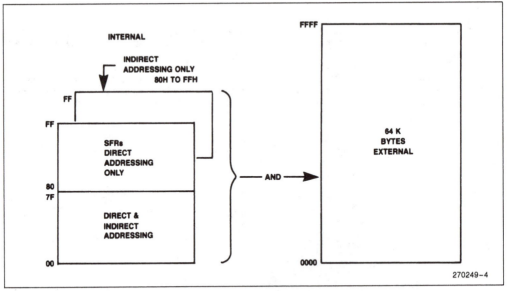

Figure 3b. The 8052 Data Memory

## INDIRECT ADDRESS AREA:

Note that in Figure 3b the SFRs and the indirect address RAM have the same addresses (80H–0FFH). Nevertheless, they are two separate areas and are accessed in two different ways.

For example the instruction

        MOV     80H, #0AAH

writes 0AAH to Port 0 which is one of the SFRs and the instruction

        MOV     R0, #80H

        MOV     @R0, #0BBH

writes 0BBH in location 80H of the data RAM. Thus, after execution of both of the above instructions Port 0 will contain 0AAH and location 80 of the RAM will contain 0BBH.

Note that the stack operations are examples of indirect addressing, so the upper 128 bytes of data RAM are available as stack space in those devices which implement 256 bytes of internal RAM.

## DIRECT AND INDIRECT ADDRESS AREA:

The 128 bytes of RAM which can be accessed by both direct and indirect addressing can be divided into 3 segments as listed below and shown in Figure 4.

**1. Register Banks 0-3:** Locations 0 through 1FH (32 bytes). ASM-51 and the device after reset default to register bank 0. To use the other register banks the user must select them in the software (refer to the MCS-51 Micro Assembler User's Guide). Each register bank contains 8 one-byte registers, 0 through 7.

Reset initializes the Stack Pointer to location 07H and it is incremented once to start from location 08H which is the first register (R0) of the second register bank. Thus, in order to use more than one register bank, the SP should be intialized to a different location of the RAM where it is not used for data storage (ie, higher part of the RAM).

**2. Bit Addressable Area:** 16 bytes have been assigned for this segment, 20H-2FH. Each one of the 128 bits of this segment can be directly addressed (0-7FH).

The bits can be referred to in two ways both of which are acceptable by the ASM-51. One way is to refer to their addresses, ie. 0 to 7FH. The other way is with reference to bytes 20H to 2FH. Thus, bits 0–7 can also be referred to as bits 20.0–20.7, and bits 8-FH are the same as 21.0–21.7 and so on.

Each of the 16 bytes in this segment can also be addressed as a byte.

**3. Scratch Pad Area:** Bytes 30H through 7FH are available to the user as data RAM. However, if the stack pointer has been initialized to this area, enough number of bytes should be left aside to prevent SP data destruction.

Figure 4 shows the different segments of the on-chip RAM.

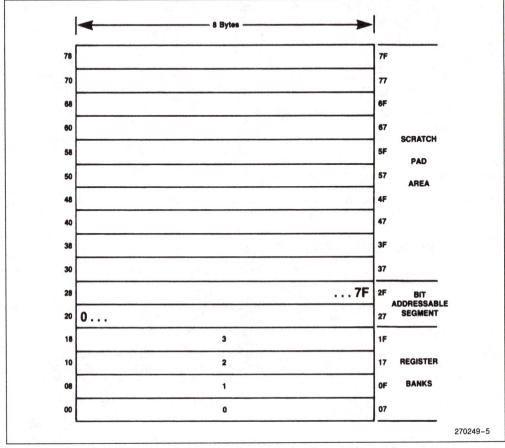

Figure 4. 128 Bytes of RAM Direct and Indirect Addressable

## SPECIAL FUNCTION REGISTERS:

Table 1 contains a list of all the SFRs and their addresses.

Comparing Table 1 and Figure 5 shows that all of the SFRs that are byte and bit addressable are located on the first column of the diagram in Figure 5.

### Table 1

| Symbol | Name | Address |
|--------|------|---------|
| *ACC | Accumulator | 0E0H |
| *B | B Register | 0F0H |
| *PSW | Program Status Word | 0D0H |
| SP | Stack Pointer | 81H |
| DPTR | Data Pointer 2 Bytes | |
|     DPL | Low Byte | 82H |
|     DPH | High Byte | 83H |
| *P0 | Port 0 | 80H |
| *P1 | Port 1 | 90H |
| *P2 | Port 2 | 0A0H |
| *P3 | Port 3 | 0B0H |
| *IP | Interrupt Priority Control | 0B8H |
| *IE | Interrupt Enable Control | 0A8H |
| TMOD | Timer/Counter Mode Control | 89H |
| *TCON | Timer/Counter Control | 88H |
| *+T2CON | Timer/Counter 2 Control | 0C8H |
| TH0 | Timer/Counter 0 High Byte | 8CH |
| TL0 | Timer/Counter 0 Low Byte | 8AH |
| TH1 | Timer/Counter 1 High Byte | 8DH |
| TL1 | Timer/Counter 1 Low Byte | 8BH |
| +TH2 | Timer/Counter 2 High Byte | 0CDH |
| +TL2 | Timer/Counter 2 Low Byte | 0CCH |
| +RCAP2H | T/C 2 Capture Reg. High Byte | 0CBH |
| +RCAP2L | T/C 2 Capture Reg. Low Byte | 0CAH |
| *SCON | Serial Control | 98H |
| SBUF | Serial Data Buffer | 99H |
| PCON | Power Control | 87H |

\* = Bit addressable
+ = 8052 only

## WHAT DO THE SFRs CONTAIN JUST AFTER POWER-ON OR A RESET?

Table 2 lists the contents of each SFR after power-on or a hardware reset.

### Table 2. Contents of the SFRs after reset

| Register | Value in Binary |
|---|---|
| *ACC | 00000000 |
| *B | 00000000 |
| *PSW | 00000000 |
| SP | 00000111 |
| DPTR | |
|     DPH | 00000000 |
|     DPL | 00000000 |
| *P0 | 11111111 |
| *P1 | 11111111 |
| *P2 | 11111111 |
| *P3 | 11111111 |
| *IP | 8051 XXX00000, |
| | 8052 XX000000 |
| *IE | 8051 0XX00000, |
| | 8052 0X000000 |
| TMOD | 00000000 |
| *TCON | 00000000 |
| *+T2CON | 00000000 |
| TH0 | 00000000 |
| TL0 | 00000000 |
| TH1 | 00000000 |
| TL1 | 00000000 |
| +TH2 | 00000000 |
| +TL2 | 00000000 |
| +RCAP2H | 00000000 |
| +RCAP2L | 00000000 |
| *SCON | 00000000 |
| SBUF | Indeterminate |
| PCON | HMOS 0XXXXXXX |
| | CHMOS 0XXX0000 |

X = Undefined
* = Bit Addressable
+ = 8052 only

## SFR MEMORY MAP

8 Bytes

| | | | | | | | | |
|---|---|---|---|---|---|---|---|---|
| F8 | | | | | | | | FF |
| F0 | B | | | | | | | F7 |
| E8 | | | | | | | | EF |
| E0 | ACC | | | | | | | E7 |
| D8 | | | | | | | | DF |
| D0 | PSW | | | | | | | D7 |
| C8 | T2CON | | RCAP2L | RCAP2H | TL2 | TH2 | | CF |
| C0 | | | | | | | | C7 |
| B8 | IP | | | | | | | BF |
| B0 | P3 | | | | | | | B7 |
| A8 | IE | | | | | | | AF |
| A0 | P2 | | | | | | | A7 |
| 98 | SCON | SBUF | | | | | | 9F |
| 90 | P1 | | | | | | | 97 |
| 88 | TCON | TMOD | TL0 | TL1 | TH0 | TH1 | | 8F |
| 80 | P0 | SP | DPL | DPH | | | | PCON | 87 |

↑
Bit
Addressable

**Figure 5**

195

Those SFRs that have their bits assigned for various functions are listed in this section. A brief description of each bit is provided for quick reference. For more detailed information refer to the Architecture Chapter of this book.

## PSW: PROGRAM STATUS WORD. BIT ADDRESSABLE.

| CY | AC | F0 | RS1 | RS0 | OV | — | P |
|----|----|----|-----|-----|----|----|----|

| | | |
|----|--------|---|
| CY | PSW.7 | Carry Flag. |
| AC | PSW.6 | Auxiliary Carry Flag. |
| F0 | PSW.5 | Flag 0 available to the user for general purpose. |
| RS1 | PSW.4 | Register Bank selector bit 1 (SEE NOTE 1). |
| RS0 | PSW.3 | Register Bank selector bit 0 (SEE NOTE 1). |
| OV | PSW.2 | Overflow Flag. |
| — | PSW.1 | User definable flag. |
| P | PSW.0 | Parity flag. Set/cleared by hardware each instruction cycle to indicate an odd/even number of '1' bits in the accumulator. |

**NOTE:**
1. The value presented by RS0 and RS1 selects the corresponding register bank.

| RS1 | RS0 | Register Bank | Address |
|-----|-----|---------------|---------|
| 0 | 0 | 0 | 00H-07H |
| 0 | 1 | 1 | 08H-0FH |
| 1 | 0 | 2 | 10H-17H |
| 1 | 1 | 3 | 18H-1FH |

## PCON: POWER CONTROL REGISTER. NOT BIT ADDRESSABLE.

| SMOD | — | — | — | GF1 | GF0 | PD | IDL |
|------|----|----|----|-----|-----|----|----|

| | |
|------|---|
| SMOD | Double baud rate bit. If Timer 1 is used to generate baud rate and SMOD = 1, the baud rate is doubled when the Serial Port is used in modes 1, 2, or 3. |
| — | Not implemented, reserved for future use.* |
| — | Not implemented, reserved for future use.* |
| — | Not implemented, reserved for future use.* |
| GF1 | General purpose flag bit. |
| GF0 | General purpose flag bit. |
| PD | Power Down bit. Setting this bit activates Power Down operation in the 80C51BH. (Available only in CHMOS). |
| IDL | Idle Mode bit. Setting this bit activates Idle Mode operation in the 80C51BH. (Available only in CHMOS). |

If 1s are written to PD and IDL at the same time, PD takes precedence.

*User software should not write 1s to reserved bits. These bits may be used in future MCS-51 products to invoke new features. In that case, the reset or inactive value of the new bit will be 0, and its active value will be 1.

196

## INTERRUPTS:

In order to use any of the interrupts in the MCS-51, the following three steps must be taken.

1. Set the EA (enable all) bit in the IE register to 1.
2. Set the corresponding individual interrupt enable bit in the IE register to 1.
3. Begin the interrupt service routine at the corresponding Vector Address of that interrupt. See Table below.

| Interrupt Source | Vector Address |
|---|---|
| IE0 | 0003H |
| TF0 | 000BH |
| IE1 | 0013H |
| TF1 | 001BH |
| RI & TI | 0023H |
| TF2 & EXF2 | 002BH |

In addition, for external interrupts, pins $\overline{INT0}$ and $\overline{INT1}$ (P3.2 and P3.3) must be set to 1, and depending on whether the interrupt is to be level or transition activated, bits IT0 or IT1 in the TCON register may need to be set to 1.

ITx = 0 level activated

ITx = 1 transition activated

## IE: INTERRUPT ENABLE REGISTER. BIT ADDRESSABLE.

If the bit is 0, the corresponding interrupt is disabled. If the bit is 1, the corresponding interrupt is enabled.

| EA | — | ET2 | ES | ET1 | EX1 | ET0 | EX0 |
|---|---|---|---|---|---|---|---|

| | | |
|---|---|---|
| EA | IE.7 | Disables all interrupts. If EA = 0, no interrupt will be acknowledged. If EA = 1, each interrupt source is individually enabled or disabled by setting or clearing its enable bit. |
| — | IE.6 | Not implemented, reserved for future use.* |
| ET2 | IE.5 | Enable or disable the Timer 2 overflow or capture interrupt (8052 only). |
| ES | IE.4 | Enable or disable the serial port interrupt. |
| ET1 | IE.3 | Enable or disable the Timer 1 overflow interrupt. |
| EX1 | IE.2 | Enable or disable External Interrupt 1. |
| ET0 | IE.1 | Enable or disable the Timer 0 overflow interrupt. |
| EX0 | IE.0 | Enable or disable External Interrupt 0. |

*User software should not write 1s to reserved bits. These bits may be used in future MCS-51 products to invoke new features. In that case, the reset or inactive value of the new bit will be 0, and its active value will be 1.

## ASSIGNING HIGHER PRIORITY TO ONE OR MORE INTERRUPTS:

In order to assign higher priority to an interrupt the corresponding bit in the IP register must be set to 1.

Remember that while an interrupt service is in progress, it cannot be interrupted by a lower or same level interrupt.

## PRIORITY WITHIN LEVEL:

Priority within level is only to resolve simultaneous requests of the same priority level.

From high to low, interrupt sources are listed below:

IE0
TF0
IE1
TF1
RI or TI
TF2 or EXF2

## IP: INTERRUPT PRIORITY REGISTER. BIT ADDRESSABLE.

If the bit is 0, the corresponding interrupt has a lower priority and if the bit is 1 the corresponding interrupt has a higher priority.

| — | — | PT2 | PS | PT1 | PX1 | PT0 | PX0 |
|---|---|-----|----|----|-----|-----|-----|

| | | |
|---|---|---|
| — | IP. 7 | Not implemented, reserved for future use.* |
| — | IP. 6 | Not implemented, reserved for future use.* |
| PT2 | IP. 5 | Defines the Timer 2 interrupt priority level (8052 only). |
| PS | IP. 4 | Defines the Serial Port interrupt priority level. |
| PT1 | IP. 3 | Defines the Timer 1 interrupt priority level. |
| PX1 | IP. 2 | Defines External Interrupt 1 priority level. |
| PT0 | IP. 1 | Defines the Timer 0 interrupt priority level. |
| PX0 | IP. 0 | Defines the External Interrupt 0 priority level. |

*User software should not write 1s to reserved bits. These bits may be used in future MCS-51 products to invoke new features. In that case, the reset or inactive value of the new bit will be 0, and its active value will be 1.

## TCON: TIMER/COUNTER CONTROL REGISTER. BIT ADDRESSABLE.

| TF1 | TR1 | TF0 | TR0 | IE1 | IT1 | IE0 | IT0 |
|-----|-----|-----|-----|-----|-----|-----|-----|

TF1    TCON. 7   Timer 1 overflow flag. Set by hardware when the Timer/Counter 1 overflows. Cleared by hardware as processor vectors to the interrupt service routine.

TR1    TCON. 6   Timer 1 run control bit. Set/cleared by software to turn Timer/Counter 1 ON/OFF.

TF0    TCON. 5   Timer 0 overflow flag. Set by hardware when the Timer/Counter 0 overflows. Cleared by hardware as processor vectors to the service routine.

TR0    TCON. 4   Timer 0 run control bit. Set/cleared by software to turn Timer/Counter 0 ON/OFF.

IE1    TCON. 3   External Interrupt 1 edge flag. Set by hardware when External Interrupt edge is detected. Cleared by hardware when interrupt is processed.

IT1    TCON. 2   Interrupt 1 type control bit. Set/cleared by software to specify falling edge/low level triggered External Interrupt.

IE0    TCON. 1   External Interrupt 0 edge flag. Set by hardware when External Interrupt edge detected. Cleared by hardware when interrupt is processed.

IT0    TCON. 0   Interrupt 0 type control bit. Set/cleared by software to specify falling edge/low level triggered External Interrupt.

## TMOD: TIMER/COUNTER MODE CONTROL REGISTER. NOT BIT ADDRESSABLE.

| GATE | C/T̄ | M1 | M0 | GATE | C/T̄ | M1 | M0 |
|------|------|----|----|------|------|----|----|

TIMER 1                    TIMER 0

GATE   When TRx (in TCON) is set and GATE = 1, TIMER/COUNTERx will run only while INTx pin is high (hardware control). When GATE = 0, TIMER/COUNTERx will run only while TRx = 1 (software control).

C/T̄    Timer or Counter selector. Cleared for Timer operation (input from internal system clock). Set for Counter operation (input from Tx input pin).

M1     Mode selector bit. (NOTE 1)

M0     Mode selector bit. (NOTE 1)

**NOTE 1:**

| M1 | M0 | | Operating Mode |
|----|----|----|----------------|
| 0 | 0 | 0 | 13-bit Timer (MCS-48 compatible) |
| 0 | 1 | 1 | 16-bit Timer/Counter |
| 1 | 0 | 2 | 8-bit Auto-Reload Timer/Counter |
| 1 | 1 | 3 | (Timer 0) TL0 is an 8-bit Timer/Counter controlled by the standard Timer 0 control bits, TH0 is an 8-bit Timer and is controlled by Timer 1 control bits. |
| 1 | 1 | 3 | (Timer 1) Timer/Counter 1 stopped. |

## TIMER SET-UP

Tables 3 through 6 give some values for TMOD which can be used to set up Timer 0 in different modes.

It is assumed that only one timer is being used at a time. If it is desired to run Timers 0 and 1 simultaneously, in any mode, the value in TMOD for Timer 0 must be ORed with the value shown for Timer 1 (Tables 5 and 6).

For example, if it is desired to run Timer 0 in mode 1 GATE (external control), and Timer 1 in mode 2 COUNTER, then the value that must be loaded into TMOD is 69H (09H from Table 3 ORed with 60H from Table 6).

Moreover, it is assumed that the user, at this point, is not ready to turn the timers on and will do that at a different point in the program by setting bit TRx (in TCON) to 1.

## TIMER/COUNTER 0

### As a Timer:

**Table 3**

| MODE | TIMER 0 FUNCTION | TMOD | |
| --- | --- | --- | --- |
| | | INTERNAL CONTROL (NOTE 1) | EXTERNAL CONTROL (NOTE 2) |
| 0 | 13-bit Timer | 00H | 08H |
| 1 | 16-bit Timer | 01H | 09H |
| 2 | 8-bit Auto-Reload | 02H | 0AH |
| 3 | two 8-bit Timers | 03H | 0BH |

### As a Counter:

**Table 4**

| MODE | COUNTER 0 FUNCTION | TMOD | |
| --- | --- | --- | --- |
| | | INTERNAL CONTROL (NOTE 1) | EXTERNAL CONTROL (NOTE 2) |
| 0 | 13-bit Timer | 04H | 0CH |
| 1 | 16-bit Timer | 05H | 0DH |
| 2 | 8-bit Auto-Reload | 06H | 0EH |
| 3 | one 8-bit Counter | 07H | 0FH |

**NOTES:**
1. The Timer is turned ON/OFF by setting/clearing bit TR0 in the software.
2. The Timer is turned ON/OFF by the 1 to 0 transition on $\overline{INT0}$ (P3.2) when TR0 = 1 (hardware control).

## TIMER/COUNTER 1

### As a Timer:

**Table 5**

| MODE | TIMER 1 FUNCTION | TMOD | |
| | | INTERNAL CONTROL (NOTE 1) | EXTERNAL CONTROL (NOTE 2) |
|---|---|---|---|
| 0 | 13-bit Timer | 00H | 80H |
| 1 | 16-bit Timer | 10H | 90H |
| 2 | 8-bit Auto-Reload | 20H | A0H |
| 3 | does not run | 30H | B0H |

### As a Counter:

**Table 6**

| MODE | COUNTER 1 FUNCTION | TMOD | |
| | | INTERNAL CONTROL (NOTE 1) | EXTERNAL CONTROL (NOTE 2) |
|---|---|---|---|
| 0 | 13-bit Timer | 40H | C0H |
| 1 | 16-bit Timer | 50H | D0H |
| 2 | 8-bit Auto-Reload | 60H | E0H |
| 3 | not available | — | — |

**NOTES:**
1. The Timer is turned ON/OFF by setting/clearing bit TR1 in the software.
2. The Timer is turned ON/OFF by the 1 to 0 transition on $\overline{INT1}$ (P3.3) when TR1 = 1 (hardware control).

## T2CON: TIMER/COUNTER 2 CONTROL REGISTER. BIT ADDRESSABLE

### 8052 Only

| TF2 | EXF2 | RCLK | TCLK | EXEN2 | TR2 | C/$\overline{T2}$ | CP/$\overline{RL2}$ |
|-----|------|------|------|-------|-----|------|------|

TF2     T2CON. 7   Timer 2 overflow flag set by hardware and cleared by software. TF2 cannot be set when either RCLK = 1 or CLK = 1

EXF2    T2CON. 6   Timer 2 external flag set when either a capture or reload is caused by a negative transition on T2EX, and EXEN2 = 1. When Timer 2 interrupt is enabled, EXF2 = 1 will cause the CPU to vector to the Timer 2 interrupt routine. EXF2 must be cleared by software.

RCLK    T2CON. 5   Receive clock flag. When set, causes the Serial Port to use Timer 2 overflow pulses for its receive clock in modes 1 & 3. RCLK = 0 causes Timer 1 overflow to be used for the receive clock.

TLCK    T2CON. 4   Transmit clock flag. When set, causes the Serial Port to use Timer 2 overflow pulses for its transmit clock in modes 1 & 3. TCLK = 0 causes Timer 1 overflows to be used for the transmit clock.

EXEN2   T2CON. 3   Timer 2 external enable flag. When set, allows a capture or reload to occur as a result of negative transition on T2EX if Timer 2 is not being used to clock the Serial Port. EXEN2 = 0 causes Timer 2 to ignore events at T2EX.

TR2     T2CON. 2   Software START/STOP control for Timer 2. A logic 1 starts the Timer.

C/$\overline{T2}$     T2CON. 1   Timer or Counter select.

                     0 = Internal Timer. 1 = External Event Counter (falling edge triggered).

CP/$\overline{RL2}$   T2CON. 0   Capture/Reload flag. When set, captures will occur on negative transitions at T2EX if EXEN2 = 1. When cleared, Auto-Reloads will occur either with Timer 2 overflows or negative transitions at T2EX when EXEN2 = 1. When either RCLK = 1 or TCLK = 1, this bit is ignored and the Timer is forced to Auto-Reload on Timer 2 overflow.

## TIMER/COUNTER 2 SET-UP

Except for the baud rate generator mode, the values given for T2CON do not include the setting of the TR2 bit. Therefore, bit TR2 must be set, separately, to turn the Timer on.

### As a Timer:

**Table 7**

| MODE | T2CON | |
|---|---|---|
| | INTERNAL CONTROL (NOTE 1) | EXTERNAL CONTROL (NOTE 2) |
| 16-bit Auto-Reload | 00H | 08H |
| 16-bit Capture | 01H | 09H |
| BAUD rate generator receive & transmit same baud rate | 34H | 36H |
| receive only | 24H | 26H |
| transmit only | 14H | 16H |

### As a Counter:

**Table 8**

| MODE | TMOD | |
|---|---|---|
| | INTERNAL CONTROL (NOTE 1) | EXTERNAL CONTROL (NOTE 2) |
| 16-bit Auto-Reload | 02H | 0AH |
| 16-bit Capture | 03H | 0BH |

**NOTES:**
1. Capture/Reload occurs only on Timer/Counter overflow.
2. Capture/Reload occurs on Timer/Counter overflow and a 1 to 0 transition on T2EX (P1.1) pin except when Timer 2 is used in the baud rate generating mode.

## SCON: SERIAL PORT CONTROL REGISTER. BIT ADDRESSABLE.

| SM0 | SM1 | SM2 | REN | TB8 | RB8 | TI | RI |
|-----|-----|-----|-----|-----|-----|----|----|

SM0     SCON. 7   Serial Port mode specifier. (NOTE 1).

SM1     SCON. 6   Serial Port mode specifier. (NOTE 1).

SM2     SCON. 5   Enables the multiprocessor communication feature in modes 2 & 3. In mode 2 or 3, if SM2 is set to 1 then RI will not be activated if the received 9th data bit (RB8) is 0. In mode 1, if SM2 = 1 then RI will not be activated if a valid stop bit was not received. In mode 0, SM2 should be 0. (See Table 9).

REN     SCON. 4   Set/Cleared by software to Enable/Disable reception.

TB8     SCON. 3   The 9th bit that will be transmitted in modes 2 & 3. Set/Cleared by software.

RB8     SCON. 2   In modes 2 & 3, is the 9th data bit that was received. In mode 1, if SM2 = 0, RB8 is the stop bit that was received. In mode 0, RB8 is not used.

TI     SCON. 1   Transmit interrupt flag. Set by hardware at the end of the 8th bit time in mode 0, or at the beginning of the stop bit in the other modes. Must be cleared by software.

RI     SCON. 0   Receive interrupt flag. Set by hardware at the end of the 8th bit time in mode 0, or halfway through the stop bit time in the other modes (except see SM2). Must be cleared by software.

**NOTE 1:**

| SM0 | SM1 | Mode | Description | Baud Rate |
|-----|-----|------|-------------|-----------|
| 0 | 0 | 0 | SHIFT REGISTER | Fosc./12 |
| 0 | 1 | 1 | 8-Bit UART | Variable |
| 1 | 0 | 2 | 9-Bit UART | Fosc./64 OR Fosc./32 |
| 1 | 1 | 3 | 9-Bit UART | Variable |

## SERIAL PORT SET-UP:

**Table 9**

| MODE | SCON | SM2 VARIATION |
|------|------|---------------|
| 0 | 10H | Single Processor Environment (SM2 = 0) |
| 1 | 50H | |
| 2 | 90H | |
| 3 | D0H | |
| 0 | NA | Multiprocessor Environment (SM2 = 1) |
| 1 | 70H | |
| 2 | B0H | |
| 3 | F0H | |

## GENERATING BAUD RATES

### Serial Port in Mode 0:

Mode 0 has a fixed baud rate which is 1/12 of the oscillator frequency. To run the serial port in this mode none of the Timer/Counters need to be set up. Only the SCON register needs to be defined.

$$\text{Baud Rate} = \frac{\text{Osc Freq}}{12}$$

### Serial Port in Mode 1:

Mode 1 has a variable baud rate. The baud rate can be generated by either Timer 1 or Timer 2 (8052 only).

## USING TIMER/COUNTER 1 TO GENERATE BAUD RATES:

For this purpose, Timer 1 is used in mode 2 (Auto-Reload). Refer to Timer Setup section of this chapter.

$$\text{Baud Rate} = \frac{\text{K x Oscillator Freq.}}{32 \times 12 \times [256 - (\text{TH1})]}$$

If SMOD = 0, then K = 1.
If SMOD = 1, then K = 2. (SMOD is the PCON register).

Most of the time the user knows the baud rate and needs to know the reload value for TH1.
Therefore, the equation to calculate TH1 can be written as:

$$\text{TH1} = 256 - \frac{\text{K x Osc Freq.}}{384 \times \text{baud rate}}$$

TH1 must be an integer value. Rounding off TH1 to the nearest integer may not produce the desired baud rate. In this case, the user may have to choose another crystal frequency.

Since the PCON register is not bit addressable, one way to set the bit is logical ORing the PCON register. (ie, ORL PCON, #80H). The address of PCON is 87H.

## USING TIMER/COUNTER 2 TO GENERATE BAUD RATES:

For this purpose, Timer 2 must be used in the baud rate generating mode. Refer to Timer 2 Setup Table in this chapter. If Timer 2 is being clocked through pin T2 (P1.0) the baud rate is:

$$\text{Baud Rate} = \frac{\text{Timer 2 Overflow Rate}}{16}$$

And if it is being clocked internally the baud rate is:

$$\text{Baud Rate} = \frac{\text{Osc Freq}}{32 \times [65536 - (\text{RCAP2H, RCAP2L})]}$$

To obtain the reload value for RCAP2H and RCAP2L the above equation can be rewritten as:

$$\text{RCAP2H, RCAP2L} = 65536 - \frac{\text{Osc Freq}}{32 \times \text{Baud Rate}}$$

## SERIAL PORT IN MODE 2:

The baud rate is fixed in this mode and is $\frac{1}{32}$ or $\frac{1}{64}$ of the oscillator frequency depending on the value of the SMOD bit in the PCON register.

In this mode none of the Timers are used and the clock comes from the internal phase 2 clock.

SMOD = 1, Baud Rate = $\frac{1}{32}$ Osc Freq.

SMOD = 0, Baud Rate = $\frac{1}{64}$ Osc Freq.

To set the SMOD bit: ORL     PCON, #80H. The address of PCON is 87H.

## SERIAL PORT IN MODE 3:

The baud rate in mode 3 is variable and sets up exactly the same as in mode 1.

# MCS®-51 INSTRUCTION SET

### Table 10. 8051 Instruction Set Summary

Interrupt Response Time: Refer to Hardware Description Chapter.

### Instructions that Affect Flag Settings[1]

| Instruction | Flag | | | Instruction | Flag | | |
|---|---|---|---|---|---|---|---|
| | C | OV | AC | | C | OV | AC |
| ADD | X | X | X | CLR C | O | | |
| ADDC | X | X | X | CPL C | X | | |
| SUBB | X | X | X | ANL C,bit | X | | |
| MUL | O | X | | ANL C,/bit | X | | |
| DIV | O | X | | ORL C,bit | X | | |
| DA | X | | | ORL C,bit | X | | |
| RRC | X | | | MOV C,bit | X | | |
| RLC | X | | | CJNE | X | | |
| SETB C | 1 | | | | | | |

[1]Note that operations on SFR byte address 208 or bit addresses 209-215 (i.e., the PSW or bits in the PSW) will also affect flag settings.

**Note on instruction set and addressing modes:**

Rn — Register R7–R0 of the currently selected Register Bank.

direct — 8-bit internal data location's address. This could be an Internal Data RAM location (0–127) or a SFR [i.e., I/O port, control register, status register, etc. (128–255)].

@Ri — 8-bit internal data RAM location (0–255) addressed indirectly through register R1 or R0.

#data — 8-bit constant included in instruction.

#data 16 — 16-bit constant included in instruction.

addr 16 — 16-bit destination address. Used by LCALL & LJMP. A branch can be anywhere within the 64K-byte Program Memory address space.

addr 11 — 11-bit destination address. Used by ACALL & AJMP. The branch will be within the same 2K-byte page of program memory as the first byte of the following instruction.

rel — Signed (two's complement) 8-bit offset byte. Used by SJMP and all conditional jumps. Range is −128 to +127 bytes relative to first byte of the following instruction.

bit — Direct Addressed bit in Internal Data RAM or Special Function Register.

| Mnemonic | | Description | Byte | Oscillator Period |
|---|---|---|---|---|
| **ARITHMETIC OPERATIONS** | | | | |
| ADD | A,Rn | Add register to Accumulator | 1 | 12 |
| ADD | A,direct | Add direct byte to Accumulator | 2 | 12 |
| ADD | A,@Ri | Add indirect RAM to Accumulator | 1 | 12 |
| ADD | A,#data | Add immediate data to Accumulator | 2 | 12 |
| ADDC | A,Rn | Add register to Accumulator with Carry | 1 | 12 |
| ADDC | A,direct | Add direct byte to Accumulator with Carry | 2 | 12 |
| ADDC | A,@Ri | Add indirect RAM to Accumulator with Carry | 1 | 12 |
| ADDC | A,#data | Add immediate data to Acc with Carry | 2 | 12 |
| SUBB | A,Rn | Subtract Register from Acc with borrow | 1 | 12 |
| SUBB | A,direct | Subtract direct byte from Acc with borrow | 2 | 12 |
| SUBB | A,@Ri | Subtract indirect RAM from ACC with borrow | 1 | 12 |
| SUBB | A,#data | Subtract immediate data from Acc with borrow | 2 | 12 |
| INC | A | Increment Accumulator | 1 | 12 |
| INC | Rn | Increment register | 1 | 12 |
| INC | direct | Increment direct byte | 2 | 12 |
| INC | @Ri | Increment direct RAM | 1 | 12 |
| DEC | A | Decrement Accumulator | 1 | 12 |
| DEC | Rn | Decrement Register | 1 | 12 |
| DEC | direct | Decrement direct byte | 2 | 12 |
| DEC | @Ri | Decrement indirect RAM | 1 | 12 |

All mnemonics copyrighted © Intel Corporation 1980

## Table 10. 8051 Instruction Set Summary (Continued)

| Mnemonic | | Description | Byte | Oscillator Period |
|---|---|---|---|---|
| **ARITHMETIC OPERATIONS** (Continued) | | | | |
| INC | DPTR | Increment Data Pointer | 1 | 24 |
| MUL | AB | Multiply A & B | 1 | 48 |
| DIV | AB | Divide A by B | 1 | 48 |
| DA | A | Decimal Adjust Accumulator | 1 | 12 |
| **LOGICAL OPERATIONS** | | | | |
| ANL | A,Rn | AND Register to Accumulator | 1 | 12 |
| ANL | A,direct | AND direct byte to Accumulator | 2 | 12 |
| ANL | A,@Ri | AND indirect RAM to Accumulator | 1 | 12 |
| ANL | A,#data | AND immediate data to Accumulator | 2 | 12 |
| ANL | direct,A | AND Accumulator to direct byte | 2 | 12 |
| ANL | direct,#data | AND immediate data to direct byte | 3 | 24 |
| ORL | A,Rn | OR register to Accumulator | 1 | 12 |
| ORL | A,direct | OR direct byte to Accumulator | 2 | 12 |
| ORL | A,@Ri | OR indirect RAM to Accumulator | 1 | 12 |
| ORL | A,#data | OR immediate data to Accumulator | 2 | 12 |
| ORL | direct,A | OR Accumulator to direct byte | 2 | 12 |
| ORL | direct,#data | OR immediate data to direct byte | 3 | 24 |
| XRL | A,Rn | Exclusive-OR register to Accumulator | 1 | 12 |
| XRL | A,direct | Exclusive-OR direct byte to Accumulator | 2 | 12 |
| XRL | A,@Ri | Exclusive-OR indirect RAM to Accumulator | 1 | 12 |
| XRL | A,#data | Exclusive-OR immediate data to Accumulator | 2 | 12 |
| XRL | direct,A | Exclusive-OR Accumulator to direct byte | 2 | 12 |
| XRL | direct,#data | Exclusive-OR immediate data to direct byte | 3 | 24 |
| CLR | A | Clear Accumulator | 1 | 12 |
| CPL | A | Complement Accumulator | 1 | 12 |

| Mnemonic | | Description | Byte | Oscillator Period |
|---|---|---|---|---|
| **LOGICAL OPERATIONS** (Continued) | | | | |
| RL | A | Rotate Accumulator Left | 1 | 12 |
| RLC | A | Rotate Accumulator Left through the Carry | 1 | 12 |
| RR | A | Rotate Accumulator Right | 1 | 12 |
| RRC | A | Rotate Accumulator Right through the Carry | 1 | 12 |
| SWAP | A | Swap nibbles within the Accumulator | 1 | 12 |
| **DATA TRANSFER** | | | | |
| MOV | A,Rn | Move register to Accumulator | 1 | 12 |
| MOV | A,direct | Move direct byte to Accumulator | 2 | 12 |
| MOV | A,@Ri | Move indirect RAM to Accumulator | 1 | 12 |
| MOV | A,#data | Move immediate data to Accumulator | 2 | 12 |
| MOV | Rn,A | Move Accumulator to register | 1 | 12 |
| MOV | Rn,direct | Move direct byte to register | 2 | 24 |
| MOV | Rn,#data | Move immediate data to register | 2 | 12 |
| MOV | direct,A | Move Accumulator to direct byte | 2 | 12 |
| MOV | direct,Rn | Move register to direct byte | 2 | 24 |
| MOV | direct,direct | Move direct byte to direct | 3 | 24 |
| MOV | direct,@Ri | Move indirect RAM to direct byte | 2 | 24 |
| MOV | direct,#data | Move immediate data to direct byte | 3 | 24 |
| MOV | @Ri,A | Move Accumulator to indirect RAM | 1 | 12 |

All mnemonics copyrighted ©Intel Corporation 1980

**Table 10. 8051 Instruction Set Summary** (Continued)

| Mnemonic | | Description | Byte | Oscillator Period |
|---|---|---|---|---|
| **DATA TRANSFER** (Continued) | | | | |
| MOV | @Ri,direct | Move direct byte to indirect RAM | 2 | 24 |
| MOV | @Ri, # data | Move immediate data to indirect RAM | 2 | 12 |
| MOV | DPTR, # data16 | Load Data Pointer with a 16-bit constant | 3 | 24 |
| MOVC | A,@A + DPTR | Move Code byte relative to DPTR to Acc | 1 | 24 |
| MOVC | A,@A + PC | Move Code byte relative to PC to Acc | 1 | 24 |
| MOVX | A,@Ri | Move External RAM (8-bit addr) to Acc | 1 | 24 |
| MOVX | A,@DPTR | Move External RAM (16-bit addr) to Acc | 1 | 24 |
| MOVX | @Ri,A | Move Acc to External RAM (8-bit addr) | 1 | 24 |
| MOVX | @DPTR,A | Move Acc to External RAM (16-bit addr) | 1 | 24 |
| PUSH | direct | Push direct byte onto stack | 2 | 24 |
| POP | direct | Pop direct byte from stack | 2 | 24 |
| XCH | A,Rn | Exchange register with Accumulator | 1 | 12 |
| XCH | A,direct | Exchange direct byte with Accumulator | 2 | 12 |
| XCH | A,@Ri | Exchange indirect RAM with Accumulator | 1 | 12 |
| XCHD | A,@Ri | Exchange low-order Digit indirect RAM with Acc | 1 | 12 |

| Mnemonic | | Description | Byte | Oscillator Period |
|---|---|---|---|---|
| **BOOLEAN VARIABLE MANIPULATION** | | | | |
| CLR | C | Clear Carry | 1 | 12 |
| CLR | bit | Clear direct bit | 2 | 12 |
| SETB | C | Set Carry | 1 | 12 |
| SETB | bit | Set direct bit | 2 | 12 |
| CPL | C | Complement Carry | 1 | 12 |
| CPL | bit | Complement direct bit | 2 | 12 |
| ANL | C,bit | AND direct bit to CARRY | 2 | 24 |
| ANL | C,/bit | AND complement of direct bit to Carry | 2 | 24 |
| ORL | C,bit | OR direct bit to Carry | 2 | 24 |
| ORL | C,/bit | OR complement of direct bit to Carry | 2 | 24 |
| MOV | C,bit | Move direct bit to Carry | 2 | 12 |
| MOV | bit,C | Move Carry to direct bit | 2 | 24 |
| JC | rel | Jump if Carry is set | 2 | 24 |
| JNC | rel | Jump if Carry not set | 2 | 24 |
| JB | bit,rel | Jump if direct Bit is set | 3 | 24 |
| JNB | bit,rel | Jump if direct Bit is Not set | 3 | 24 |
| JBC | bit,rel | Jump if direct Bit is set & clear bit | 3 | 24 |
| **PROGRAM BRANCHING** | | | | |
| ACALL | addr11 | Absolute Subroutine Call | 2 | 24 |
| LCALL | addr16 | Long Subroutine Call | 3 | 24 |
| RET | | Return from Subroutine | 1 | 24 |
| RETI | | Return from interrupt | 1 | 24 |
| AJMP | addr11 | Absolute Jump | 2 | 24 |
| LJMP | addr16 | Long Jump | 3 | 24 |
| SJMP | rel | Short Jump (relative addr) | 2 | 24 |

All mnemonics copyrighted © Intel Corporation 1980

**Table 10. 8051 Instruction Set Summary** (Continued)

| Mnemonic | | Description | Byte | Oscillator Period |
|---|---|---|---|---|
| **PROGRAM BRANCHING** (Continued) | | | | |
| JMP | @A+DPTR | Jump indirect relative to the DPTR | 1 | 24 |
| JZ | rel | Jump if Accumulator is Zero | 2 | 24 |
| JNZ | rel | Jump if Accumulator is Not Zero | 2 | 24 |
| CJNE | A,direct,rel | Compare direct byte to Acc and Jump if Not Equal | 3 | 24 |
| CJNE | A,#data,rel | Compare immediate to Acc and Jump if Not Equal | 3 | 24 |

| Mnemonic | | Description | Byte | Oscillator Period |
|---|---|---|---|---|
| **PROGRAM BRANCHING** (Continued) | | | | |
| CJNE | Rn,#data,rel | Compare immediate to register and Jump if Not Equal | 3 | 24 |
| CJNE | @Ri,#data,rel | Compare immediate to indirect and Jump if Not Equal | 3 | 24 |
| DJNZ | Rn,rel | Decrement register and Jump if Not Zero | 2 | 24 |
| DJNZ | direct,rel | Decrement direct byte and Jump if Not Zero | 3 | 24 |
| NOP | | No Operation | 1 | 12 |

All mnemonics copyrighted © Intel Corporation 1980

### Table 11. Instruction Opcodes in Hexadecimal Order

| Hex Code | Number of Bytes | Mnemonic | Operands | Hex Code | Number of Bytes | Mnemonic | Operands |
|---|---|---|---|---|---|---|---|
| 00 | 1 | NOP | | 33 | 1 | RLC | A |
| 01 | 2 | AJMP | code addr | 34 | 2 | ADDC | A, # data |
| 02 | 3 | LJMP | code addr | 35 | 2 | ADDC | A,data addr |
| 03 | 1 | RR | A | 36 | 1 | ADDC | A,@R0 |
| 04 | 1 | INC | A | 37 | 1 | ADDC | A,@R1 |
| 05 | 2 | INC | data addr | 38 | 1 | ADDC | A,R0 |
| 06 | 1 | INC | @R0 | 39 | 1 | ADDC | A,R1 |
| 07 | 1 | INC | @R1 | 3A | 1 | ADDC | A,R2 |
| 08 | 1 | INC | R0 | 3B | 1 | ADDC | A,R3 |
| 09 | 1 | INC | R1 | 3C | 1 | ADDC | A,R4 |
| 0A | 1 | INC | R2 | 3D | 1 | ADDC | A,R5 |
| 0B | 1 | INC | R3 | 3E | 1 | ADDC | A,R6 |
| 0C | 1 | INC | R4 | 3F | 1 | ADDC | A,R7 |
| 0D | 1 | INC | R5 | 40 | 2 | JC | code addr |
| 0E | 1 | INC | R6 | 41 | 2 | AJMP | code addr |
| 0F | 1 | INC | R7 | 42 | 2 | ORL | data addr,A |
| 10 | 3 | JBC | bit addr, code addr | 43 | 3 | ORL | data addr, # data |
| 11 | 2 | ACALL | code addr | 44 | 2 | ORL | A, # data |
| 12 | 3 | LCALL | code addr | 45 | 2 | ORL | A,data addr |
| 13 | 1 | RRC | A | 46 | 1 | ORL | A,@R0 |
| 14 | 1 | DEC | A | 47 | 1 | ORL | A,@R1 |
| 15 | 2 | DEC | data addr | 48 | 1 | ORL | A,R0 |
| 16 | 1 | DEC | @R0 | 49 | 1 | ORL | A,R1 |
| 17 | 1 | DEC | @R1 | 4A | 1 | ORL | A,R2 |
| 18 | 1 | DEC | R0 | 4B | 1 | ORL | A,R3 |
| 19 | 1 | DEC | R1 | 4C | 1 | ORL | A,R4 |
| 1A | 1 | DEC | R2 | 4D | 1 | ORL | A,R5 |
| 1B | 1 | DEC | R3 | 4E | 1 | ORL | A,R6 |
| 1C | 1 | DEC | R4 | 4F | 1 | ORL | A,R7 |
| 1D | 1 | DEC | R5 | 50 | 2 | JNC | code addr |
| 1E | 1 | DEC | R6 | 51 | 2 | ACALL | code addr |
| 1F | 1 | DEC | R7 | 52 | 2 | ANL | data addr,A |
| 20 | 3 | JB | bit addr, code addr | 53 | 3 | ANL | data addr, # data |
| 21 | 2 | AJMP | code addr | 54 | 2 | ANL | A, # data |
| 22 | 1 | RET | | 55 | 2 | ANL | A,data addr |
| 23 | 1 | RL | A | 56 | 1 | ANL | A,@R0 |
| 24 | 2 | ADD | A, # data | 57 | 1 | ANL | A,@R1 |
| 25 | 2 | ADD | A,data addr | 58 | 1 | ANL | A,R0 |
| 26 | 1 | ADD | A,@R0 | 59 | 1 | ANL | A,R1 |
| 27 | 1 | ADD | A,@R1 | 5A | 1 | ANL | A,R2 |
| 28 | 1 | ADD | A,R0 | 5B | 1 | ANL | A,R3 |
| 29 | 1 | ADD | A,R1 | 5C | 1 | ANL | A,R4 |
| 2A | 1 | ADD | A,R2 | 5D | 1 | ANL | A,R5 |
| 2B | 1 | ADD | A,R3 | 5E | 1 | ANL | A,R6 |
| 2C | 1 | ADD | A,R4 | 5F | 1 | ANL | A,R7 |
| 2D | 1 | ADD | A,R5 | 60 | 2 | JZ | code addr |
| 2E | 1 | ADD | A,R6 | 61 | 2 | AJMP | code addr |
| 2F | 1 | ADD | A,R7 | 62 | 2 | XRL | data addr,A |
| 30 | 3 | JNB | bit addr, code addr | 63 | 3 | XRL | data addr, # data |
| 31 | 2 | ACALL | code addr | 64 | 2 | XRL | A, # data |
| 32 | 1 | RETI | | 65 | 2 | XRL | A,data addr |

### Table 11. Instruction Opcodes in Hexadecimal Order (Continued)

| Hex Code | Number of Bytes | Mnemonic | Operands | Hex Code | Number of Bytes | Mnemonic | Operands |
|---|---|---|---|---|---|---|---|
| 66 | 1 | XRL | A,@R0 | 99 | 1 | SUBB | A,R1 |
| 67 | 1 | XRL | A,@R1 | 9A | 1 | SUBB | A,R2 |
| 68 | 1 | XRL | A,R0 | 9B | 1 | SUBB | A,R3 |
| 69 | 1 | XRL | A,R1 | 9C | 1 | SUBB | A,R4 |
| 6A | 1 | XRL | A,R2 | 9D | 1 | SUBB | A,R5 |
| 6B | 1 | XRL | A,R3 | 9E | 1 | SUBB | A,R6 |
| 6C | 1 | XRL | A,R4 | 9F | 1 | SUBB | A,R7 |
| 6D | 1 | XRL | A,R5 | A0 | 2 | ORL | C,/bit addr |
| 6E | 1 | XRL | A,R6 | A1 | 2 | AJMP | code addr |
| 6F | 1 | XRL | A,R7 | A2 | 2 | MOV | C,bit addr |
| 70 | 2 | JNZ | code addr | A3 | 1 | INC | DPTR |
| 71 | 2 | ACALL | code addr | A4 | 1 | MUL | AB |
| 72 | 2 | ORL | C,bit addr | A5 |  | reserved |  |
| 73 | 1 | JMP | @A+DPTR | A6 | 2 | MOV | @R0,data addr |
| 74 | 2 | MOV | A,#data | A7 | 2 | MOV | @R1,data addr |
| 75 | 3 | MOV | data addr,#data | A8 | 2 | MOV | R0,data addr |
| 76 | 2 | MOV | @R0,#data | A9 | 2 | MOV | R1,data addr |
| 77 | 2 | MOV | @R1,#data | AA | 2 | MOV | R2,data addr |
| 78 | 2 | MOV | R0,#data | AB | 2 | MOV | R3,data addr |
| 79 | 2 | MOV | R1,#data | AC | 2 | MOV | R4,data addr |
| 7A | 2 | MOV | R2,#data | AD | 2 | MOV | R5,data addr |
| 7B | 2 | MOV | R3,#data | AE | 2 | MOV | R6,data addr |
| 7C | 2 | MOV | R4,#data | AF | 2 | MOV | R7,data addr |
| 7D | 2 | MOV | R5,#data | B0 | 2 | ANL | C,/bit addr |
| 7E | 2 | MOV | R6,#data | B1 | 2 | ACALL | code addr |
| 7F | 2 | MOV | R7,#data | B2 | 2 | CPL | bit addr |
| 80 | 2 | SJMP | code addr | B3 | 1 | CPL | C |
| 81 | 2 | AJMP | code addr | B4 | 3 | CJNE | A,#data,code addr |
| 82 | 2 | ANL | C,bit addr | B5 | 3 | CJNE | A,data addr,code addr |
| 83 | 1 | MOVC | A,@A+PC | B6 | 3 | CJNE | @R0,#data,code addr |
| 84 | 1 | DIV | AB | B7 | 3 | CJNE | @R1,#data,code addr |
| 85 | 3 | MOV | data addr, data addr | B8 | 3 | CJNE | R0,#data,code addr |
| 86 | 2 | MOV | data addr,@R0 | B9 | 3 | CJNE | R1,#data,code addr |
| 87 | 2 | MOV | data addr,@R1 | BA | 3 | CJNE | R2,#data,code addr |
| 88 | 2 | MOV | data addr,R0 | BB | 3 | CJNE | R3,#data,code addr |
| 89 | 2 | MOV | data addr,R1 | BC | 3 | CJNE | R4,#data,code addr |
| 8A | 2 | MOV | data addr,R2 | BD | 3 | CJNE | R5,#data,code addr |
| 8B | 2 | MOV | data addr,R3 | BE | 3 | CJNE | R6,#data,code addr |
| 8C | 2 | MOV | data addr,R4 | BF | 3 | CJNE | R7,#data,code addr |
| 8D | 2 | MOV | data addr,R5 | C0 | 2 | PUSH | data addr |
| 8E | 2 | MOV | data addr,R6 | C1 | 2 | AJMP | code addr |
| 8F | 2 | MOV | data addr,R7 | C2 | 2 | CLR | bit addr |
| 90 | 3 | MOV | DPTR,#data | C3 | 1 | CLR | C |
| 91 | 2 | ACALL | code addr | C4 | 1 | SWAP | A |
| 92 | 2 | MOV | bit addr,C | C5 | 2 | XCH | A,data addr |
| 93 | 1 | MOVC | A,@A+DPTR | C6 | 1 | XCH | A,@R0 |
| 94 | 2 | SUBB | A,#data | C7 | 1 | XCH | A,@R1 |
| 95 | 2 | SUBB | A,data addr | C8 | 1 | XCH | A,R0 |
| 96 | 1 | SUBB | A,@R0 | C9 | 1 | XCH | A,R1 |
| 97 | 1 | SUBB | A,@R1 | CA | 1 | XCH | A,R2 |
| 98 | 1 | SUBB | A,R0 | CB | 1 | XCH | A,R3 |

### Table 11. Instruction Opcodes in Hexadecimal Order (Continued)

| Hex Code | Number of Bytes | Mnemonic | Operands | Hex Code | Number of Bytes | Mnemonic | Operands |
|---|---|---|---|---|---|---|---|
| CC | 1 | XCH | A,R4 | E6 | 1 | MOV | A,@R0 |
| CD | 1 | XCH | A,R5 | E7 | 1 | MOV | A,@R1 |
| CE | 1 | XCH | A,R6 | E8 | 1 | MOV | A,R0 |
| CF | 1 | XCH | A,R7 | E9 | 1 | MOV | A,R1 |
| D0 | 2 | POP | data addr | EA | 1 | MOV | A,R2 |
| D1 | 2 | ACALL | code addr | EB | 1 | MOV | A,R3 |
| D2 | 2 | SETB | bit addr | EC | 1 | MOV | A,R4 |
| D3 | 1 | SETB | C | ED | 1 | MOV | A,R5 |
| D4 | 1 | DA | A | EE | 1 | MOV | A,R6 |
| D5 | 3 | DJNZ | data addr,code addr | EF | 1 | MOV | A,R7 |
| D6 | 1 | XCHD | A,@R0 | F0 | 1 | MOVX | @DPTR,A |
| D7 | 1 | XCHD | A,@R1 | F1 | 2 | ACALL | code addr |
| D8 | 2 | DJNZ | R0,code addr | F2 | 1 | MOVX | @R0,A |
| D9 | 2 | DJNZ | R1,code addr | F3 | 1 | MOVX | @R1,A |
| DA | 2 | DJNZ | R2,code addr | F4 | 1 | CPL | A |
| DB | 2 | DJNZ | R3,code addr | F5 | 2 | MOV | data addr,A |
| DC | 2 | DJNZ | R4,code addr | F6 | 1 | MOV | @R0,A |
| DD | 2 | DJNZ | R5,code addr | F7 | 1 | MOV | @R1,A |
| DE | 2 | DJNZ | R6,code addr | F8 | 1 | MOV | R0,A |
| DF | 2 | DJNZ | R7,code addr | F9 | 1 | MOV | R1,A |
| E0 | 1 | MOVX | A,@DPTR | FA | 1 | MOV | R2,A |
| E1 | 2 | AJMP | code addr | FB | 1 | MOV | R3,A |
| E2 | 1 | MOVX | A,@R0 | FC | 1 | MOV | R4,A |
| E3 | 1 | MOVX | A,@R1 | FD | 1 | MOV | R5,A |
| E4 | 1 | CLR | A | FE | 1 | MOV | R6,A |
| E5 | 2 | MOV | A,data addr | FF | 1 | MOV | R7,A |

# INSTRUCTION DEFINITIONS

## ACALL   addr11

| | |
|---|---|
| **Function:** | Absolute Call |
| **Description:** | ACALL unconditionally calls a subroutine located at the indicated address. The instruction increments the PC twice to obtain the address of the following instruction, then pushes the 16-bit result onto the stack (low-order byte first) and increments the Stack Pointer twice. The destination address is obtained by successively concatenating the five high-order bits of the incremented PC, opcode bits 7-5, and the second byte of the instruction. The subroutine called must therefore start within the same 2K block of the program memory as the first byte of the instruction following ACALL. No flags are affected. |
| **Example:** | Initially SP equals 07H. The label "SUBRTN" is at program memory location 0345 H. After executing the instruction,

ACALL   SUBRTN

at location 0123H, SP will contain 09H, internal RAM locations 08H and 09H will contain 25H and 01H, respectively, and the PC will contain 0345H. |
| **Bytes:** | 2 |
| **Cycles:** | 2 |

**Encoding:**

| a10 a9 a8 1 | 0 0 0 1 | | a7 a6 a5 a4 | a3 a2 a1 a0 |
|---|---|---|---|---|

**Operation:**   ACALL

$(PC) \leftarrow (PC) + 2$
$(SP) \leftarrow (SP) + 1$
$((SP)) \leftarrow (PC_{7-0})$
$(SP) \leftarrow (SP) + 1$
$((SP)) \leftarrow (PC_{15-8})$
$(PC_{10-0}) \leftarrow$ page address

## ADD   A,<src-byte>

**Function:**   Add

**Description:**   ADD adds the byte variable indicated to the Accumulator, leaving the result in the Accumulator. The carry and auxiliary-carry flags are set, respectively, if there is a carry-out from bit 7 or bit 3, and cleared otherwise. When adding unsigned integers, the carry flag indicates an overflow occured.

OV is set if there is a carry-out of bit 6 but not out of bit 7, or a carry-out of bit 7 but not bit 6; otherwise OV is cleared. When adding signed integers, OV indicates a negative number produced as the sum of two positive operands, or a positive sum from two negative operands.

Four source operand addressing modes are allowed: register, direct, register-indirect, or immediate.

**Example:**   The Accumulator holds 0C3H (11000011B) and register 0 holds 0AAH (10101010B). The instruction,

ADD   A,R0

will leave 6DH (01101101B) in the Accumulator with the AC flag cleared and both the carry flag and OV set to 1.

## ADD   A,Rn

**Bytes:**   1

**Cycles:**   1

**Encoding:**

| 0 0 1 0 | 1 r r r |
|---------|---------|

**Operation:**   ADD
(A) ← (A) + (Rn)

## ADD   A,direct

**Bytes:**   2

**Cycles:**   1

**Encoding:**

| 0 0 1 0 | 0 1 0 1 | | direct address |
|---------|---------|---|----------------|

**Operation:**   ADD
(A) ← (A) + (direct)

**ADD   A,@Ri**

    **Bytes:**   1

    **Cycles:**   1

    **Encoding:**

| 0 0 1 0 | 0 1 1 i |
|---|---|

    **Operation:**   ADD
                  $(A) \leftarrow (A) + ((R_i))$

**ADD   A,#data**

    **Bytes:**   2

    **Cycles:**   1

    **Encoding:**

| 0 0 1 0 | 0 1 0 0 | immediate data |
|---|---|---|

    **Operation:**   ADD
                  $(A) \leftarrow (A) + \#data$

## ADDC   A,<src-byte>

    **Function:**   Add with Carry

    **Description:**   ADDC simultaneously adds the byte variable indicated, the carry flag and the Accumulator contents, leaving the result in the Accumulator. The carry and auxiliary-carry flags are set, respectively, if there is a carry-out from bit 7 or bit 3, and cleared otherwise. When adding unsigned integers, the carry flag indicates an overflow occured.

                  OV is set if there is a carry-out of bit 6 but not out of bit 7, or a carry-out of bit 7 but not out of bit 6; otherwise OV is cleared. When adding signed integers, OV indicates a negative number produced as the sum of two positive operands or a positive sum from two negative operands.

                  Four source operand addressing modes are allowed: register, direct, register-indirect, or immediate.

    **Example:**   The Accumulator holds 0C3H (11000011B) and register 0 holds 0AAH (10101010B) with the carry flag set. The instruction,

                  ADDC   A,R0

                  will leave 6EH (01101110B) in the Accumulator with AC cleared and both the Carry flag and OV set to 1.

**ADDC   A,Rn**

      **Bytes:**  1

      **Cycles:**  1

      **Encoding:**

| 0 0 1 1 | 1 r r r |
|---|---|

      **Operation:**  ADDC
                    $(A) \leftarrow (A) + (C) + (R_n)$

**ADDC   A,direct**

      **Bytes:**  2

      **Cycles:**  1

      **Encoding:**

| 0 0 1 1 | 0 1 0 1 | direct address |
|---|---|---|

      **Operation:**  ADDC
                    $(A) \leftarrow (A) + (C) + (direct)$

**ADDC   A,@Ri**

      **Bytes:**  1

      **Cycles:**  1

      **Encoding:**

| 0 0 1 1 | 0 1 1 i |
|---|---|

      **Operation:**  ADDC
                    $(A) \leftarrow (A) + (C) + ((R_i))$

**ADDC   A,#data**

      **Bytes:**  2

      **Cycles:**  1

      **Encoding:**

| 0 0 1 1 | 0 1 0 0 | immediate data |
|---|---|---|

      **Operation:**  ADDC
                    $(A) \leftarrow (A) + (C) + \#data$

## AJMP  addr11

| | |
|---|---|
| **Function:** | Absolute Jump |
| **Description:** | AJMP transfers program execution to the indicated address, which is formed at run-time by concatenating the high-order five bits of the PC (*after* incrementing the PC twice), opcode bits 7-5, and the second byte of the instruction. The destination must therefore be within the same 2K block of program memory as the first byte of the instruction following AJMP. |
| **Example:** | The label "JMPADR" is at program memory location 0123H. The instruction, |

AJMP  JMPADR

is at location 0345H and will load the PC with 0123H.

| | |
|---|---|
| **Bytes:** | 2 |
| **Cycles:** | 2 |

**Encoding:**

| a10 a9 a8 0 | 0 0 0 1 | | a7 a6 a5 a4 | a3 a2 a1 a0 |
|---|---|---|---|---|

**Operation:** AJMP
$(PC) \leftarrow (PC) + 2$
$(PC_{10-0}) \leftarrow$ page address

## ANL  <dest-byte>,<src-byte>

| | |
|---|---|
| **Function:** | Logical-AND for byte variables |
| **Description:** | ANL performs the bitwise logical-AND operation between the variables indicated and stores the results in the destination variable. No flags are affected. |

The two operands allow six addressing mode combinations. When the destination is the Accumulator, the source can use register, direct, register-indirect, or immediate addressing; when the destination is a direct address, the source can be the Accumulator or immediate data.

*Note:* When this instruction is used to modify an output port, the value used as the original port data will be read from the output data latch, *not* the input pins.

| | |
|---|---|
| **Example:** | If the Accumulator holds 0C3H (11000011B) and register 0 holds 55H (01010101B) then the instruction, |

ANL  A,R0

will leave 41H (01000001B) in the Accumulator.

When the destination is a directly addressed byte, this instruction will clear combinations of bits in any RAM location or hardware register. The mask byte determining the pattern of bits to be cleared would either be a constant contained in the instruction or a value computed in the Accumulator at run-time. The instruction,

ANL  P1,#01110011B

will clear bits 7, 3, and 2 of output port 1.

217

**ANL   A,Rn**

| | |
|---|---|
| **Bytes:** | 1 |
| **Cycles:** | 1 |

**Encoding:**

| 0 1 0 1 | 1 r r r |
|---|---|

**Operation:**   ANL
(A) ← (A) ∧ (Rn)

**ANL   A,direct**

| | |
|---|---|
| **Bytes:** | 2 |
| **Cycles:** | 1 |

**Encoding:**

| 0 1 0 1 | 0 1 0 1 | | direct address |
|---|---|---|---|

**Operation:**   ANL

(A) ← (A) ∧ (direct)

**ANL   A,@Ri**

| | |
|---|---|
| **Bytes:** | 1 |
| **Cycles:** | 1 |

**Encoding:**

| 0 1 0 1 | 0 1 1 i |
|---|---|

**Operation:**   ANL

(A) ← (A) ∧ ((Ri))

**ANL   A,#data**

| | |
|---|---|
| **Bytes:** | 2 |
| **Cycles:** | 1 |

**Encoding:**

| 0 1 0 1 | 0 1 0 0 | | immediate data |
|---|---|---|---|

**Operation:**   ANL
(A) ← (A) ∧ #data

**ANL   direct,A**

| | |
|---|---|
| **Bytes:** | 2 |
| **Cycles:** | 1 |

**Encoding:**

| 0 1 0 1 | 0 0 1 0 | | direct address |
|---|---|---|---|

**Operation:**   ANL
(direct) ← (direct) ∧ (A)

218

**ANL   direct,#data**

> **Bytes:**   3
>
> **Cycles:**   2
>
> **Encoding:**   | 0 1 0 1 | 0 0 1 1 |    | direct address |    | immediate data |
>
> **Operation:**   ANL
> (direct) ← (direct) ∧ #data

**ANL   C,<src-bit>**

> **Function:**   Logical-AND for bit variables
>
> **Description:**   If the Boolean value of the source bit is a logical 0 then clear the carry flag; otherwise leave the carry flag in its current state. A slash ("/") preceding the operand in the assembly language indicates that the logical complement of the addressed bit is used as the source value, *but the source bit itself is not affected.* No other flags are affected.
>
> Only direct addressing is allowed for the source operand.
>
> **Example:**   Set the carry flag if, and only if, P1.0 = 1, ACC. 7 = 1, and OV = 0:
>
> MOV   C,P1.0     ;LOAD CARRY WITH INPUT PIN STATE
>
> ANL   C,ACC.7  ;AND CARRY WITH ACCUM. BIT 7
>
> ANL   C,/OV      ;AND WITH INVERSE OF OVERFLOW FLAG

**ANL   C,bit**

> **Bytes:**   2
>
> **Cycles:**   2
>
> **Encoding:**   | 1 0 0 0 | 0 0 1 0 |    | bit address |
>
> **Operation:**   ANL
> (C) ← (C) ∧ (bit)

**ANL   C,/bit**

> **Bytes:**   2
>
> **Cycles:**   2
>
> **Encoding:**   | 1 0 1 1 | 0 0 0 0 |    | bit address |
>
> **Operation:**   ANL
> (C) ← (C) ∧ ⌐ (bit)

219

---

**CJNE**  <dest-byte>,<src-byte>, rel

---

**Function:**    Compare and Jump if Not Equal.

**Description:**    CJNE compares the magnitudes of the first two operands, and branches if their values are not equal. The branch destination is computed by adding the signed relative-displacement in the last instruction byte to the PC, after incrementing the PC to the start of the next instruction. The carry flag is set if the unsigned integer value of <dest-byte> is less than the unsigned integer value of <src-byte>; otherwise, the carry is cleared. Neither operand is affected.

The first two operands allow four addressing mode combinations: the Accumulator may be compared with any directly addressed byte or immediate data, and any indirect RAM location or working register can be compared with an immediate constant.

**Example:**    The Accumulator contains 34H. Register 7 contains 56H. The first instruction in the sequence,

```
            CJNE  R7,#60H, NOT__EQ
;             . . .    . . . . .           ;  R7 = 60H.
NOT__EQ       JC     REQ__LOW              ;  IF R7 < 60H.
;             . . .    . . . . .           ;  R7 > 60H.
```

sets the carry flag and branches to the instruction at label NOT__EQ. By testing the carry flag, this instruction determines whether R7 is greater or less than 60H.

If the data being presented to Port 1 is also 34H, then the instruction,

WAIT:  CJNE  A,P1,WAIT

clears the carry flag and continues with the next instruction in sequence, since the Accumulator does equal the data read from P1. (If some other value was being input on P1, the program will loop at this point until the P1 data changes to 34H.)

---

**CJNE   A,direct,rel**

**Bytes:**    3

**Cycles:**    2

**Encoding:**    | 1 0 1 1 | 0 1 0 1 |    | direct address |    | rel. address |

**Operation:**    $(PC) \leftarrow (PC) + 3$
IF $(A) <> (direct)$
THEN
    $(PC) \leftarrow (PC) + relative\ offset$

IF $(A) < (direct)$
THEN
    $(C) \leftarrow 1$
ELSE
    $(C) \leftarrow 0$

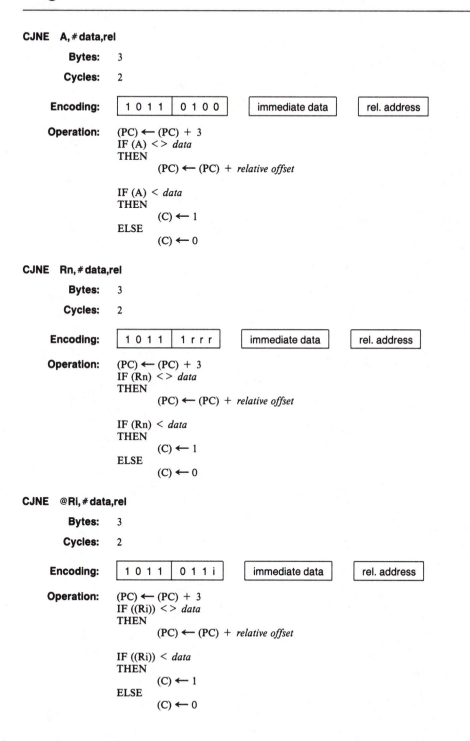

**CJNE A,#data,rel**

> **Bytes:** 3

> **Cycles:** 2

> **Encoding:** | 1 0 1 1 | 0 1 0 0 |   immediate data   |   rel. address   |

> **Operation:** (PC) ← (PC) + 3
> IF (A) <> *data*
> THEN
> > (PC) ← (PC) + *relative offset*
>
> IF (A) < *data*
> THEN
> > (C) ← 1
> ELSE
> > (C) ← 0

**CJNE Rn,#data,rel**

> **Bytes:** 3

> **Cycles:** 2

> **Encoding:** | 1 0 1 1 | 1 r r r |   immediate data   |   rel. address   |

> **Operation:** (PC) ← (PC) + 3
> IF (Rn) <> *data*
> THEN
> > (PC) ← (PC) + *relative offset*
>
> IF (Rn) < *data*
> THEN
> > (C) ← 1
> ELSE
> > (C) ← 0

**CJNE @Ri,#data,rel**

> **Bytes:** 3

> **Cycles:** 2

> **Encoding:** | 1 0 1 1 | 0 1 1 i |   immediate data   |   rel. address   |

> **Operation:** (PC) ← (PC) + 3
> IF ((Ri)) <> *data*
> THEN
> > (PC) ← (PC) + *relative offset*
>
> IF ((Ri)) < *data*
> THEN
> > (C) ← 1
> ELSE
> > (C) ← 0

## CLR A

| | |
|---|---|
| **Function:** | Clear Accumulator |
| **Description:** | The Accumulator is cleared (all bits set on zero). No flags are affected. |
| **Example:** | The Accumulator contains 5CH (01011100B). The instruction, |

CLR   A

will leave the Accumulator set to 00H (00000000B).

| | |
|---|---|
| **Bytes:** | 1 |
| **Cycles:** | 1 |
| **Encoding:** | 1 1 1 0 \| 0 1 0 0 |
| **Operation:** | CLR<br>(A) ← 0 |

## CLR bit

| | |
|---|---|
| **Function:** | Clear bit |
| **Description:** | The indicated bit is cleared (reset to zero). No other flags are affected. CLR can operate on the carry flag or any directly addressable bit. |
| **Example:** | Port 1 has previously been written with 5DH (01011101B). The instruction, |

CLR   P1.2

will leave the port set to 59H (01011001B).

## CLR C

| | |
|---|---|
| **Bytes:** | 1 |
| **Cycles:** | 1 |
| **Encoding:** | 1 1 0 0 \| 0 0 1 1 |
| **Operation:** | CLR<br>(C) ← 0 |

## CLR bit

| | |
|---|---|
| **Bytes:** | 2 |
| **Cycles:** | 1 |
| **Encoding:** | 1 1 0 0 \| 0 0 1 0    bit address |
| **Operation:** | CLR<br>(bit) ← 0 |

## CPL A

| | |
|---|---|
| **Function:** | Complement Accumulator |
| **Description:** | Each bit of the Accumulator is logically complemented (one's complement). Bits which previously contained a one are changed to a zero and vice-versa. No flags are affected. |
| **Example:** | The Accumulator contains 5CH (01011100B). The instruction, |

CPL   A

will leave the Accumulator set to 0A3H (10100011B).

| | |
|---|---|
| **Bytes:** | 1 |
| **Cycles:** | 1 |
| **Encoding:** | 1 1 1 1 \| 0 1 0 0 |
| **Operation:** | CPL<br>(A) ← ⌐ (A) |

## CPL   bit

| | |
|---|---|
| **Function:** | Complement bit |
| **Description:** | The bit variable specified is complemented. A bit which had been a one is changed to zero and vice-versa. No other flags are affected. CLR can operate on the carry or any directly addressable bit.

*Note:* When this instruction is used to modify an output pin, the value used as the original data will be read from the output data latch, *not* the input pin. |
| **Example:** | Port 1 has previously been written with 5BH (01011101B). The instruction sequence, |

CPL   P1.1

CPL   P1.2

will leave the port set to 5BH (01011011B).

## CPL   C

| | |
|---|---|
| **Bytes:** | 1 |
| **Cycles:** | 1 |
| **Encoding:** | 1 0 1 1 \| 0 0 1 1 |
| **Operation:** | CPL<br>(C) ← ⌐ (C) |

**CPL    bit**

| | |
|---|---|
| **Bytes:** | 2 |
| **Cycles:** | 1 |

**Encoding:**

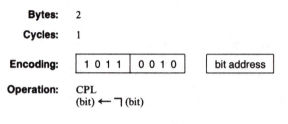

| 1 0 1 1 | 0 0 1 0 |   | bit address |

**Operation:**    CPL
(bit) ← ⌐ (bit)

---

**DA    A**

**Function:**    Decimal-adjust Accumulator for Addition

**Description:**    DA A adjusts the eight-bit value in the Accumulator resulting from the earlier addition of two variables (each in packed-BCD format), producing two four-bit digits. Any ADD or ADDC instruction may have been used to perform the addition.

If Accumulator bits 3-0 are greater than nine (xxxx1010-xxxx1111), or if the AC flag is one, six is added to the Accumulator producing the proper BCD digit in the low-order nibble. This internal addition would set the carry flag if a carry-out of the low-order four-bit field propagated through all high-order bits, but it would not clear the carry flag otherwise.

If the carry flag is now set, or if the four high-order bits now exceed nine (1010xxxx-111xxxx), these high-order bits are incremented by six, producing the proper BCD digit in the high-order nibble. Again, this would set the carry flag if there was a carry-out of the high-order bits, but wouldn't clear the carry. The carry flag thus indicates if the sum of the original two BCD variables is greater than 100, allowing multiple precision decimal addition. OV is not affected.

All of this occurs during the one instruction cycle. Essentially, this instruction performs the decimal conversion by adding 00H, 06H, 60H, or 66H to the Accumulator, depending on initial Accumulator and PSW conditions.

*Note:* DA A *cannot* simply convert a hexadecimal number in the Accumulator to BCD notation, nor does DA A apply to decimal subtraction.

**Example:** The Accumulator holds the value 56H (01010110B) representing the packed BCD digits of the decimal number 56. Register 3 contains the value 67H (01100111B) representing the packed BCD digits of the decimal number 67. The carry flag is set. The instruction sequence.

```
ADDC  A,R3
DA    A
```

will first perform a standard twos-complement binary addition, resulting in the value 0BEH (10111110) in the Accumulator. The carry and auxiliary carry flags will be cleared.

The Decimal Adjust instruction will then alter the Accumulator to the value 24H (00100100B), indicating the packed BCD digits of the decimal number 24, the low-order two digits of the decimal sum of 56, 67, and the carry-in. The carry flag will be set by the Decimal Adjust instruction, indicating that a decimal overflow occurred. The true sum 56, 67, and 1 is 124.

BCD variables can be incremented or decremented by adding 01H or 99H. If the Accumulator initially holds 30H (representing the digits of 30 decimal), then the instruction sequence,

```
ADD   A,#99H

DA    A
```

will leave the carry set and 29H in the Accumulator, since $30 + 99 = 129$. The low-order byte of the sum can be interpreted to mean $30 - 1 = 29$.

**Bytes:** 1

**Cycles:** 1

**Encoding:**

| 1 1 0 1 | 0 1 0 0 |
|---------|---------|

**Operation:** DA
-contents of Accumulator are BCD
IF    $[[(A_{3-0}) > 9] \vee [(AC) = 1]]$
        THEN$(A_{3-0}) \leftarrow (A_{3-0}) + 6$
                AND

IF    $[[(A_{7-4}) > 9] \vee [(C) = 1]]$
        THEN $(A_{7-4}) \leftarrow (A_{7-4}) + 6$

## DEC  byte

**Function:** Decrement

**Description:** The variable indicated is decremented by 1. An original value of 00H will underflow to 0FFH. No flags are affected. Four operand addressing modes are allowed: accumulator, register, direct, or register-indirect.

*Note:* When this instruction is used to modify an output port, the value used as the original port data will be read from the output data latch, *not* the input pins.

**Example:** Register 0 contains 7FH (01111111B). Internal RAM locations 7EH and 7FH contain 00H and 40H, respectively. The instruction sequence,

DEC  @R0

DEC  R0

DEC  @R0

will leave register 0 set to 7EH and internal RAM locations 7EH and 7FH set to 0FFH and 3FH.

## DEC  A

**Bytes:** 1

**Cycles:** 1

**Encoding:**

| 0 0 0 1 | 0 1 0 0 |
|---------|---------|

**Operation:** DEC
(A) ← (A) − 1

## DEC  Rn

**Bytes:** 1

**Cycles:** 1

**Encoding:**

| 0 0 0 1 | 1 r r r |
|---------|---------|

**Operation:** DEC
(Rn) ← (Rn) − 1

**DEC   direct**

> **Bytes:** 2
>
> **Cycles:** 1
>
> **Encoding:**    | 0 0 0 1 | 0 1 0 1 |    | direct address |
>
> **Operation:** DEC
> (direct) ← (direct) − 1

**DEC   @Ri**

> **Bytes:** 1
>
> **Cycles:** 1
>
> **Encoding:**    | 0 0 0 1 | 0 1 1 i |
>
> **Operation:** DEC
> ((Ri)) ← ((Ri)) − 1

---

**DIV   AB**

> **Function:** Divide
>
> **Description:** DIV AB divides the unsigned eight-bit integer in the Accumulator by the unsigned eight-bit integer in register B. The Accumulator receives the integer part of the quotient; register B receives the integer remainder. The carry and OV flags will be cleared.
>
> *Exception:* if B had originally contained 00H, the values returned in the Accumulator and B-register will be undefined and the overflow flag will be set. The carry flag is cleared in any case.
>
> **Example:** The Accumulator contains 251 (0FBH or 11111011B) and B contains 18 (12H or 00010010B). The instruction,
>
> DIV   AB
>
> will leave 13 in the Accumulator (0DH or 00001101B) and the value 17 (11H or 00010001B) in B, since 251 = (13 X 18) + 17. Carry and OV will both be cleared.
>
> **Bytes:** 1
>
> **Cycles:** 4
>
> **Encoding:**    | 1 0 0 0 | 0 1 0 0 |
>
> **Operation:** DIV
> $(A)_{15-8}$
> $(B)_{7-0}$ ← $(A)/(B)$

## DJNZ  <byte>,<rel-addr>

**Function:**     Decrement and Jump if Not Zero

**Description:**     DJNZ decrements the location indicated by 1, and branches to the address indicated by the second operand if the resulting value is not zero. An original value of 00H will underflow to 0FFH. No flags are affected. The branch destination would be computed by adding the signed relative-displacement value in the last instruction byte to the PC, after incrementing the PC to the first byte of the following instruction.

The location decremented may be a register or directly addressed byte.

*Note:* When this instruction is used to modify an output port, the value used as the original port data will be read from the output data latch, *not* the input pins.

**Example:**     Internal RAM locations 40H, 50H, and 60H contain the values 01H, 70H, and 15H, respectively. The instruction sequence,

```
DJNZ   40H,LABEL_1
DJNZ   50H,LABEL_2
DJNZ   60H,LABEL_3
```

will cause a jump to the instruction at label LABEL_2 with the values 00H, 6FH, and 15H in the three RAM locations. The first jump was *not* taken because the result was zero.

This instruction provides a simple way of executing a program loop a given number of times, or for adding a moderate time delay (from 2 to 512 machine cycles) with a single instruction. The instruction sequence,

```
                MOV      R2,#8
TOGGLE:         CPL      P1.7
                DJNZ     R2,TOGGLE
```

will toggle P1.7 eight times, causing four output pulses to appear at bit 7 of output Port 1. Each pulse will last three machine cycles; two for DJNZ and one to alter the pin.

## DJNZ  Rn,rel

**Bytes:**     2

**Cycles:**     2

**Encoding:**

| 1 1 0 1 | 1 r r r | | rel. address |
|---------|---------|--|--------------|

**Operation:**     DJNZ
$(PC) \leftarrow (PC) + 2$
$(Rn) \leftarrow (Rn) - 1$
IF   $(Rn) > 0$ or $(Rn) < 0$
     THEN
            $(PC) \leftarrow (PC) + rel$

**DJNZ    direct,rel**

**Bytes:** 3

**Cycles:** 2

**Encoding:**  | 1 1 0 1 | 0 1 0 1 |    | direct address |    | rel. address |

**Operation:**    DJNZ
$(PC) \leftarrow (PC) + 2$
$(direct) \leftarrow (direct) - 1$
IF $(direct) > 0$ or $(direct) < 0$
    THEN
        $(PC) \leftarrow (PC) + rel$

---

**INC    <byte>**

**Function:**    Increment

**Description:**    INC increments the indicated variable by 1. An original value of 0FFH will overflow to 00H. No flags are affected. Three addressing modes are allowed: register, direct, or register-indirect.

*Note:* When this instruction is used to modify an output port, the value used as the original port data will be read from the output data latch, *not* the input pins.

**Example:**    Register 0 contains 7EH (01111110B). Internal RAM locations 7EH and 7FH contain 0FFH and 40H, respectively. The instruction sequence,

    INC   @R0
    INC   R0
    INC   @R0

will leave register 0 set to 7FH and internal RAM locations 7EH and 7FH holding (respectively) 00H and 41H.

**INC    A**

**Bytes:** 1

**Cycles:** 1

**Encoding:**  | 0 0 0 0 | 0 1 0 0 |

**Operation:**    INC
$(A) \leftarrow (A) + 1$

## INC   Rn

      **Bytes:**    1

      **Cycles:**    1

    **Encoding:**

| 0 0 0 0 | 1 r r r |
|---------|---------|

   **Operation:**    INC
                     (Rn) ← (Rn) + 1

## INC   direct

      **Bytes:**    2

      **Cycles:**    1

    **Encoding:**

| 0 0 0 0 | 0 1 0 1 | direct address |
|---------|---------|----------------|

   **Operation:**    INC
                     (direct) ← (direct) + 1

## INC   @Ri

      **Bytes:**    1

      **Cycles:**    1

    **Encoding:**

| 0 0 0 0 | 0 1 1 i |
|---------|---------|

   **Operation:**    INC
                     ((Ri)) ← ((Ri)) + 1

## INC   DPTR

   **Function:**    Increment Data Pointer

 **Description:**    Increment the 16-bit data pointer by 1. A 16-bit increment (modulo $2^{16}$) is performed; an overflow of the low-order byte of the data pointer (DPL) from 0FFH to 00H will increment the high-order byte (DPH). No flags are affected.

                This is the only 16-bit register which can be incremented.

     **Example:**    Registers DPH and DPL contain 12H and 0FEH, respectively. The instruction sequence,

```
INC   DPTR
INC   DPTR
INC   DPTR
```

                will change DPH and DPL to 13H and 01H.

      **Bytes:**    1

      **Cycles:**    2

    **Encoding:**

| 1 0 1 0 | 0 0 1 1 |
|---------|---------|

   **Operation:**    INC
                     (DPTR) ← (DPTR) + 1

## JB   bit,rel

**Function:** Jump if Bit set

**Description:** If the indicated bit is a one, jump to the address indicated; otherwise proceed with the next instruction. The branch destination is computed by adding the signed relative-displacement in the third instruction byte to the PC, after incrementing the PC to the first byte of the next instruction. *The bit tested is not modified.* No flags are affected.

**Example:** The data present at input port 1 is 11001010B. The Accumulator holds 56 (01010110B). The instruction sequence,

JB   P1.2,LABEL1

JB   ACC.2,LABEL2

will cause program execution to branch to the instruction at label LABEL2.

**Bytes:** 3

**Cycles:** 2

**Encoding:**

| 0 0 1 0 | 0 0 0 0 | | bit address | | rel. address |
|---|---|---|---|---|---|

**Operation:** JB
$$(PC) \leftarrow (PC) + 3$$
IF   (bit) = 1
    THEN
$$(PC) \leftarrow (PC) + rel$$

## JBC   bit,rel

**Function:** Jump if Bit is set and Clear bit

**Description:** If the indicated bit is one, branch to the address indicated; otherwise proceed with the next instruction. *The bit will not be cleared if it is already a zero.* The branch destination is computed by adding the signed relative-displacement in the third instruction byte to the PC, after incrementing the PC to the first byte of the next instruction. No flags are affected.

*Note:* When this instruction is used to test an output pin, the value used as the original data will be read from the output data latch, *not* the input pin.

**Example:** The Accumulator holds 56H (01010110B). The instruction sequence,

JBC   ACC.3,LABEL1
JBC   ACC.2,LABEL2

will cause program execution to continue at the instruction identified by the label LABEL2, with the Accumulator modified to 52H (01010010B).

**Bytes:** 3

**Cycles:** 2

**Encoding:**

| 0 0 0 1 | 0 0 0 0 | | bit address | | rel. address |

**Operation:** JBC
$(PC) \leftarrow (PC) + 3$
IF (bit) = 1
    THEN
        (bit) $\leftarrow$ 0
        $(PC) \leftarrow (PC)$ + rel

---

## JC rel

---

**Function:** Jump if Carry is set

**Description:** If the carry flag is set, branch to the address indicated; otherwise proceed with the next instruction. The branch destination is computed by adding the signed relative-displacement in the second instruction byte to the PC, after incrementing the PC twice. No flags are affected.

**Example:** The carry flag is cleared. The instruction sequence,

```
JC      LABEL1
CPL     C
JC      LABEL 2
```

will set the carry and cause program execution to continue at the instruction identified by the label LABEL2.

**Bytes:** 2

**Cycles:** 2

**Encoding:**

| 0 1 0 0 | 0 0 0 0 | | rel. address |

**Operation:** JC
$(PC) \leftarrow (PC) + 2$
IF (C) = 1
    THEN
        $(PC) \leftarrow (PC)$ + rel

**JMP   @A+DPTR**

| | |
|---|---|
| **Function:** | Jump indirect |
| **Description:** | Add the eight-bit unsigned contents of the Accumulator with the sixteen-bit data pointer, and load the resulting sum to the program counter. This will be the address for subsequent instruction fetches. Sixteen-bit addition is performed (modulo $2^{16}$): a carry-out from the low-order eight bits propagates through the higher-order bits. Neither the Accumulator nor the Data Pointer is altered. No flags are affected. |
| **Example:** | An even number from 0 to 6 is in the Accumulator. The following sequence of instructions will branch to one of four AJMP instructions in a jump table starting at JMP__TBL: |

```
                MOV     DPTR,#JMP__TBL
                JMP     @A+DPTR
JMP__TBL:       AJMP    LABEL0
                AJMP    LABEL1
                AJMP    LABEL2
                AJMP    LABEL3
```

If the Accumulator equals 04H when starting this sequence, execution will jump to label LABEL2. Remember that AJMP is a two-byte instruction, so the jump instructions start at every other address.

| | |
|---|---|
| **Bytes:** | 1 |
| **Cycles:** | 2 |
| **Encoding:** | 0 1 1 1    0 0 1 1 |
| **Operation:** | JMP<br>(PC) ← (A) + (DPTR) |

## JNB  bit,rel

| | |
|---|---|
| **Function:** | Jump if Bit Not set |

**Description:** If the indicated bit is a zero, branch to the indicated address; otherwise proceed with the next instruction. The branch destination is computed by adding the signed relative-displacement in the third instruction byte to the PC, after incrementing the PC to the first byte of the next instruction. *The bit tested is not modified*. No flags are affected.

**Example:** The data present at input port 1 is 11001010B. The Accumulator holds 56H (01010110B). The instruction sequence,

```
JNB   P1.3,LABEL1
JNB   ACC.3,LABEL2
```

will cause program execution to continue at the instruction at label LABEL2.

**Bytes:** 3

**Cycles:** 2

**Encoding:** 
| 0 0 1 1 | 0 0 0 0 | bit address | rel. address |
|---|---|---|---|

**Operation:** JNB  
(PC) ← (PC) + 3  
IF  (bit) = 0  
    THEN (PC) ← (PC) + rel.

## JNC  rel

**Function:** Jump if Carry not set

**Description:** If the carry flag is a zero, branch to the address indicated; otherwise proceed with the next instruction. The branch destination is computed by adding the signed relative-displacement in the second instruction byte to the PC, after incrementing the PC twice to point to the next instruction. The carry flag is not modified.

**Example:** The carry flag is set. The instruction sequence,

```
JNC   LABEL1
CPL   C
JNC   LABEL2
```

will clear the carry and cause program execution to continue at the instruction identified by the label LABEL2.

**Bytes:** 2

**Cycles:** 2

**Encoding:** 
| 0 1 0 1 | 0 0 0 0 | rel. address |
|---|---|---|

**Operation:** JNC  
(PC) ← (PC) + 2  
IF  (C) = 0  
    THEN  (PC) ← (PC) + rel

234

## JNZ   rel

| | |
|---|---|
| **Function:** | Jump if Accumulator Not Zero |
| **Description:** | If any bit of the Accumulator is a one, branch to the indicated address; otherwise proceed with the next instruction. The branch destination is computed by adding the signed relative-displacement in the second instruction byte to the PC, after incrementing the PC twice. The Accumulator is not modified. No flags are affected. |
| **Example:** | The Accumulator originally holds 00H. The instruction sequence, |

```
JNZ   LABEL1
INC   A
JNZ   LABEL2
```

will set the Accumulator to 01H and continue at label LABEL2.

| | |
|---|---|
| **Bytes:** | 2 |
| **Cycles:** | 2 |
| **Encoding:** | `0 1 1 1` `0 0 0 0`    `rel. address` |
| **Operation:** | JNZ |

$$(PC) \leftarrow (PC) + 2$$
$$IF \quad (A) \neq 0$$
$$\qquad THEN \quad (PC) \leftarrow (PC) + rel$$

## JZ   rel

| | |
|---|---|
| **Function:** | Jump if Accumulator Zero |
| **Description:** | If all bits of the Accumulator are zero, branch to the address indicated; otherwise proceed with the next instruction. The branch destination is computed by adding the signed relative-displacement in the second instruction byte to the PC, after incrementing the PC twice. The Accumulator is not modified. No flags are affected. |
| **Example:** | The Accumulator originally contains 01H. The instruction sequence, |

```
JZ    LABEL1
DEC   A
JZ    LABEL2
```

will change the Accumulator to 00H and cause program execution to continue at the instruction identified by the label LABEL2.

| | |
|---|---|
| **Bytes:** | 2 |
| **Cycles:** | 2 |
| **Encoding:** | `0 1 1 0` `0 0 0 0`    `rel. address` |
| **Operation:** | JZ |

$$(PC) \leftarrow (PC) + 2$$
$$IF \quad (A) = 0$$
$$\qquad THEN \quad (PC) \leftarrow (PC) + rel$$

## LCALL addr16

**Function:** Long call

**Description:** LCALL calls a subroutine located at the indicated address. The instruction adds three to the program counter to generate the address of the next instruction and then pushes the 16-bit result onto the stack (low byte first), incrementing the Stack Pointer by two. The high-order and low-order bytes of the PC are then loaded, respectively, with the second and third bytes of the LCALL instruction. Program execution continues with the instruction at this address. The subroutine may therefore begin anywhere in the full 64K-byte program memory address space. No flags are affected.

**Example:** Initially the Stack Pointer equals 07H. The label "SUBRTN" is assigned to program memory location 1234H. After executing the instruction,

LCALL   SUBRTN

at location 0123H, the Stack Pointer will contain 09H, internal RAM locations 08H and 09H will contain 26H and 01H, and the PC will contain 1234H.

**Bytes:** 3

**Cycles:** 2

**Encoding:**

| 0 0 0 1 | 0 0 1 0 | addr15-addr8 | addr7-addr0 |
|---------|---------|--------------|-------------|

**Operation:** LCALL
$(PC) \leftarrow (PC) + 3$
$(SP) \leftarrow (SP) + 1$
$((SP)) \leftarrow (PC_{7-0})$
$(SP) \leftarrow (SP) + 1$
$((SP)) \leftarrow (PC_{15-8})$
$(PC) \leftarrow addr_{15-0}$

## LJMP addr16

**Function:** Long Jump

**Description:** LJMP causes an unconditional branch to the indicated address, by loading the high-order and low-order bytes of the PC (respectively) with the second and third instruction bytes. The destination may therefore be anywhere in the full 64K program memory address space. No flags are affected.

**Example:** The label "JMPADR" is assigned to the instruction at program memory location 1234H. The instruction,

LJMP   JMPADR

at location 0123H will load the program counter with 1234H.

**Bytes:** 3

**Cycles:** 2

**Encoding:**

| 0 0 0 0 | 0 0 1 0 | addr15-addr8 | addr7-addr0 |
|---------|---------|--------------|-------------|

**Operation:** LJMP
$(PC) \leftarrow addr_{15-0}$

236

---

**MOV** <dest-byte>,<src-byte>

---

**Function:** Move byte variable

**Description:** The byte variable indicated by the second operand is copied into the location specified by the first operand. The source byte is not affected. No other register or flag is affected.

This is by far the most flexible operation. Fifteen combinations of source and destination addressing modes are allowed.

**Example:** Internal RAM location 30H holds 40H. The value of RAM location 40H is 10H. The data present at input port 1 is 11001010B (0CAH).

```
MOV   R0,#30H   ;R0 <= 30H
MOV   A,@R0     ;A <= 40H
MOV   R1,A      ;R1 <= 40H
MOV   B,@R1     ;B <= 10H
MOV   @R1,P1    ;RAM (40H) <= 0CAH
MOV   P2,P1     ;P2 #0CAH
```

leaves the value 30H in register 0, 40H in both the Accumulator and register 1, 10H in register B, and 0CAH (11001010B) both in RAM location 40H and output on port 2.

**MOV   A,Rn**

**Bytes:** 1

**Cycles:** 1

**Encoding:**

| 1 1 1 0 | 1 r r r |
|---------|---------|

**Operation:** MOV
(A) ← (Rn)

***MOV   A,direct**

**Bytes:** 2

**Cycles:** 1

**Encoding:**

| 1 1 1 0 | 0 1 0 1 | direct address |
|---------|---------|----------------|

**Operation:** MOV
(A) ← (direct)

**MOV A,ACC is not a valid instruction.**

**MOV   A,@Ri**

    **Bytes:**   1

    **Cycles:**   1

    **Encoding:**   | 1 1 1 0 | 0 1 1 i |

    **Operation:**   MOV
                (A) ← ((Ri))

**MOV   A,#data**

    **Bytes:**   2

    **Cycles:**   1

    **Encoding:**   | 0 1 1 1 | 0 1 0 0 |   | immediate data |

    **Operation:**   MOV
                (A) ← #data

**MOV   Rn,A**

    **Bytes:**   1

    **Cycles:**   1

    **Encoding:**   | 1 1 1 1 | 1 r r r |

    **Operation:**   MOV
                (Rn) ← (A)

**MOV   Rn,direct**

    **Bytes:**   2

    **Cycles:**   2

    **Encoding:**   | 1 0 1 0 | 1 r r r |   | direct addr. |

    **Operation:**   MOV
                (Rn) ← (direct)

**MOV   Rn,#data**

    **Bytes:**   2

    **Cycles:**   1

    **Encoding:**   | 0 1 1 1 | 1 r r r |   | immediate data |

    **Operation:**   MOV
                (Rn) ← #data

**MOV   direct,A**

       **Bytes:**   2

      **Cycles:**  1

    **Encoding:**

| 1 1 1 1 | 0 1 0 1 | | direct address |
|---|---|---|---|

    **Operation:**  MOV
                 (direct) ← (A)

**MOV   direct,Rn**

       **Bytes:**   2

      **Cycles:**  2

    **Encoding:**

| 1 0 0 0 | 1 r r r | | direct address |
|---|---|---|---|

    **Operation:**  MOV
                 (direct) ← (Rn)

**MOV   direct,direct**

       **Bytes:**   3

      **Cycles:**  2

    **Encoding:**

| 1 0 0 0 | 0 1 0 1 | | dir. addr. (src) | dir. addr. (dest) |
|---|---|---|---|---|

    **Operation:**  MOV
                 (direct) ← (direct)

**MOV   direct,@Ri**

       **Bytes:**   2

      **Cycles:**  2

    **Encoding:**

| 1 0 0 0 | 0 1 1 i | | direct addr. |
|---|---|---|---|

    **Operation:**  MOV
                 (direct) ← ((Ri))

**MOV   direct,#data**

       **Bytes:**   3

      **Cycles:**  2

    **Encoding:**

| 0 1 1 1 | 0 1 0 1 | | direct address | immediate data |
|---|---|---|---|---|

    **Operation:**  MOV
                 (direct) ← #data

**MOV @Ri,A**

>           **Bytes:** 1
>
>          **Cycles:** 1
>
>        **Encoding:**
>
> | 1 1 1 1 | 0 1 1 i |
> |---|---|
>
>       **Operation:** MOV
> ((Ri)) ← (A)

**MOV @Ri,direct**

>           **Bytes:** 2
>
>          **Cycles:** 2
>
>        **Encoding:**
>
> | 1 0 1 0 | 0 1 1 i | | direct addr. |
> |---|---|---|---|
>
>       **Operation:** MOV
> ((Ri)) ← (direct)

**MOV @Ri,#data**

>           **Bytes:** 2
>
>          **Cycles:** 1
>
>        **Encoding:**
>
> | 0 1 1 1 | 0 1 1 i | | immediate data |
> |---|---|---|---|
>
>       **Operation:** MOV
> ((RI)) ← #data

---

**MOV <dest-bit>,<src-bit>**

---

>       **Function:** Move bit data
>
>    **Description:** The Boolean variable indicated by the second operand is copied into the location specified by the first operand. One of the operands must be the carry flag; the other may be any directly addressable bit. No other register or flag is affected.
>
>        **Example:** The carry flag is originally set. The data present at input Port 3 is 11000101B. The data previously written to output Port 1 is 35H (00110101B).
>
> MOV  P1.3,C
> MOV  C,P3.3
> MOV  P1.2,C
>
> will leave the carry cleared and change Port 1 to 39H (00111001B).

240

## MOV   C,bit

**Bytes:**   2

**Cycles:**   1

**Encoding:**   | 1 0 1 0 | 0 0 1 0 |   | bit address |

**Operation:**   MOV
(C) ← (bit)

## MOV   bit,C

**Bytes:**   2

**Cycles:**   2

**Encoding:**   | 1 0 0 1 | 0 0 1 0 |   | bit address |

**Operation:**   MOV
(bit) ← (C)

## MOV   DPTR,#data16

**Function:**   Load Data Pointer with a 16-bit constant

**Description:**   The Data Pointer is loaded with the 16-bit constant indicated. The 16-bit constant is loaded into the second and third bytes of the instruction. The second byte (DPH) is the high-order byte, while the third byte (DPL) holds the low-order byte. No flags are affected.

This is the only instruction which moves 16 bits of data at once.

**Example:**   The instruction,

MOV   DPTR,#1234H

will load the value 1234H into the Data Pointer: DPH will hold 12H and DPL will hold 34H.

**Bytes:**   3

**Cycles:**   2

**Encoding:**   | 1 0 0 1 | 0 0 0 0 |   | immed. data15-8 |   | immed. data7-0 |

**Operation:**   MOV
(DPTR) ← #$data_{15-0}$
DPH □ DPL ← #$data_{15-8}$ □ #$data_{7-0}$

## MOVC A,@A+ <base-reg>

| | |
|---|---|
| **Function:** | Move Code byte |
| **Description:** | The MOVC instructions load the Accumulator with a code byte, or constant from program memory. The address of the byte fetched is the sum of the original unsigned eight-bit Accumulator contents and the contents of a sixteen-bit base register, which may be either the Data Pointer or the PC. In the latter case, the PC is incremented to the address of the following instruction before being added with the Accumulator; otherwise the base register is not altered. Sixteen-bit addition is performed so a carry-out from the low-order eight bits may propagate through higher-order bits. No flags are affected. |
| **Example:** | A value between 0 and 3 is in the Accumulator. The following instructions will translate the value in the Accumulator to one of four values defined by the DB (define byte) directive. |

```
REL__PC:  INC    A

          MOVC   A,@A+PC

          RET

          DB     66H

          DB     77H

          DB     88H

          DB     99H
```

If the subroutine is called with the Accumulator equal to 01H, it will return with 77H in the Accumulator. The INC A before the MOVC instruction is needed to "get around" the RET instruction above the table. If several bytes of code separated the MOVC from the table, the corresponding number would be added to the Accumulator instead.

## MOVC A,@A+DPTR

| | |
|---|---|
| **Bytes:** | 1 |
| **Cycles:** | 2 |
| **Encoding:** | `1 0 0 1` `0 0 1 1` |
| **Operation:** | MOVC<br>(A) ← ((A) + (DPTR)) |

## MOVC A,@A + PC

| | |
|---|---|
| **Bytes:** | 1 |
| **Cycles:** | 2 |
| **Encoding:** | `1 0 0 0` `0 0 1 1` |
| **Operation:** | MOVC<br>(PC) ← (PC) + 1<br>(A) ← ((A) + (PC)) |

**MOVX    <dest-byte>,<src-byte>**

**Function:**    Move External

**Description:**    The MOVX instructions transfer data between the Accumulator and a byte of external data memory, hence the "X" appended to MOV. There are two types of instructions, differing in whether they provide an eight-bit or sixteen-bit indirect address to the external data RAM.

In the first type, the contents of R0 or R1 in the current register bank provide an eight-bit address multiplexed with data on P0. Eight bits are sufficient for external I/O expansion decoding or for a relatively small RAM array. For somewhat larger arrays, any output port pins can be used to output higher-order address bits. These pins would be controlled by an output instruction preceding the MOVX.

In the second type of MOVX instruction, the Data Pointer generates a sixteen-bit address. P2 outputs the high-order eight address bits (the contents of DPH) while P0 multiplexes the low-order eight bits (DPL) with data. The P2 Special Function Register retains its previous contents while the P2 output buffers are emitting the contents of DPH. This form is faster and more efficient when accessing very large data arrays (up to 64K bytes), since no additional instructions are needed to set up the output ports.

It is possible in some situations to mix the two MOVX types. A large RAM array with its high-order address lines driven by P2 can be addressed via the Data Pointer, or with code to output high-order address bits to P2 followed by a MOVX instruction using R0 or R1.

**Example:**    An external 256 byte RAM using multiplexed address/data lines (e.g., an Intel 8155 RAM/I/O/Timer) is connected to the 8051 Port 0. Port 3 provides control lines for the external RAM. Ports 1 and 2 are used for normal I/O. Registers 0 and 1 contain 12H and 34H. Location 34H of the external RAM holds the value 56H. The instruction sequence,

MOVX    A,@R1

MOVX    @R0,A

copies the value 56H into both the Accumulator and external RAM location 12H.

**MOVX   A,@Ri**

|  | |
|---|---|
| **Bytes:** | 1 |
| **Cycles:** | 2 |

**Encoding:**  | 1 1 1 0 | 0 0 1 i |

**Operation:**  MOVX
(A) ← ((Ri))

**MOVX   A,@DPTR**

|  | |
|---|---|
| **Bytes:** | 1 |
| **Cycles:** | 2 |

**Encoding:**  | 1 1 1 0 | 0 0 0 0 |

**Operation:**  MOVX
(A) ← ((DPTR))

**MOVX   @Ri,A**

|  | |
|---|---|
| **Bytes:** | 1 |
| **Cycles:** | 2 |

**Encoding:**  | 1 1 1 1 | 0 0 1 i |

**Operation:**  MOVX
((Ri)) ← (A)

**MOVX   @DPTR,A**

|  | |
|---|---|
| **Bytes:** | 1 |
| **Cycles:** | 2 |

**Encoding:**  | 1 1 1 1 | 0 0 0 0 |

**Operation:**  MOVX
(DPTR) ← (A)

244

## MUL AB

**Function:** Multiply

**Description:** MUL AB multiplies the unsigned eight-bit integers in the Accumulator and register B. The low-order byte of the sixteen-bit product is left in the Accumulator, and the high-order byte in B. If the product is greater than 255 (0FFH) the overflow flag is set; otherwise it is cleared. The carry flag is always cleared.

**Example:** Originally the Accumulator holds the value 80 (50H). Register B holds the value 160 (0A0H). The instruction,

MUL   AB

will give the product 12,800 (3200H), so B is changed to 32H (00110010B) and the Accumulator is cleared. The overflow flag is set, carry is cleared.

**Bytes:** 1

**Cycles:** 4

**Encoding:**

| 1 0 1 0 | 0 1 0 0 |
|---------|---------|

**Operation:** MUL
$(A)_{7-0} \leftarrow (A) \times (B)$
$(B)_{15-8}$

## NOP

**Function:** No Operation

**Description:** Execution continues at the following instruction. Other than the PC, no registers or flags are affected.

**Example:** It is desired to produce a low-going output pulse on bit 7 of Port 2 lasting exactly 5 cycles. A simple SETB/CLR sequence would generate a one-cycle pulse, so four additional cycles must be inserted. This may be done (assuming no interrupts are enabled) with the instruction sequence,

CLR     P2.7
NOP
NOP
NOP
NOP
SETB    P2.7

**Bytes:** 1

**Cycles:** 1

**Encoding:**

| 0 0 0 0 | 0 0 0 0 |
|---------|---------|

**Operation:** NOP
$(PC) \leftarrow (PC) + 1$

## ORL   \<dest-byte\> \<src-byte\>

**Function:**     Logical-OR for byte variables

**Description:**   ORL performs the bitwise logical-OR operation between the indicated variables, storing the results in the destination byte. No flags are affected.

The two operands allow six addressing mode combinations. When the destination is the Accumulator, the source can use register, direct, register-indirect, or immediate addressing; when the destination is a direct address, the source can be the Accumulator or immediate data.

*Note:* When this instruction is used to modify an output port, the value used as the original port data will be read from the output data latch, *not* the input pins.

**Example:**     If the Accumulator holds 0C3H (11000011B) and R0 holds 55H (01010101B) then the instruction,

ORL   A,R0

will leave the Accumulator holding the value 0D7H (11010111B).

When the destination is a directly addressed byte, the instruction can set combinations of bits in any RAM location or hardware register. The pattern of bits to be set is determined by a mask byte, which may be either a constant data value in the instruction or a variable computed in the Accumulator at run-time. The instruction,

ORL   P1,#00110010B

will set bits 5, 4, and 1 of output Port 1.

## ORL   A,Rn

**Bytes:**     1

**Cycles:**    1

**Encoding:**

| 0 1 0 0 | 1 r r r |
|---------|---------|

**Operation:**   ORL
$(A) \leftarrow (A) \vee (Rn)$

**ORL   A,direct**

| | |
|---|---|
| **Bytes:** | 2 |
| **Cycles:** | 1 |

**Encoding:**  | 0 1 0 0 | 0 1 0 1 |   | direct address |

**Operation:**   ORL
(A) ← (A) ∨ (direct)

**ORL   A,@Ri**

| | |
|---|---|
| **Bytes:** | 1 |
| **Cycles:** | 1 |

**Encoding:**  | 0 1 0 0 | 0 1 1 i |

**Operation:**   ORL
(A) ← (A) ∨ ((Ri))

**ORL   A,#data**

| | |
|---|---|
| **Bytes:** | 2 |
| **Cycles:** | 1 |

**Encoding:**  | 0 1 0 0 | 0 1 0 0 |   | immediate data |

**Operation:**   ORL
(A) ← (A) ∨ #data

**ORL   direct,A**

| | |
|---|---|
| **Bytes:** | 2 |
| **Cycles:** | 1 |

**Encoding:**  | 0 1 0 0 | 0 0 1 0 |   | direct address |

**Operation:**   ORL
(direct) ← (direct) ∨ (A)

**ORL   direct,#data**

| | |
|---|---|
| **Bytes:** | 3 |
| **Cycles:** | 2 |

**Encoding:**  | 0 1 0 0 | 0 0 1 1 |   | direct addr. |   | immediate data |

**Operation:**   ORL
(direct) ← (direct) ∨ #data

## ORL  C,<src-bit>

**Function:** Logical-OR for bit variables

**Description:** Set the carry flag if the Boolean value is a logical 1; leave the carry in its current state otherwise . A slash ("/") preceding the operand in the assembly language indicates that the logical complement of the addressed bit is used as the source value, but the source bit itself is not affected. No other flags are affected.

**Example:** Set the carry flag if and only if P1.0 = 1, ACC. 7 = 1, or OV = 0:

MOV  C,P1.0    ;LOAD CARRY WITH INPUT PIN P10

ORL  C,ACC.7  ;OR CARRY WITH THE ACC. BIT 7

ORL  C,/OV    ;OR CARRY WITH THE INVERSE OF OV.

## ORL  C,bit

**Bytes:** 2

**Cycles:** 2

**Encoding:**

| 0 1 1 1 | 0 0 1 0 | | bit address |
|---------|---------|--|-------------|

**Operation:** ORL
$(C) \leftarrow (C) \lor (bit)$

## ORL  C,/bit

**Bytes:** 2

**Cycles:** 2

**Encoding:**

| 1 0 1 0 | 0 0 0 0 | | bit address |
|---------|---------|--|-------------|

**Operation:** ORL
$(C) \leftarrow (C) \lor (\overline{bit})$

---

## POP direct

| | |
|---|---|
| **Function:** | Pop from stack. |
| **Description:** | The contents of the internal RAM location addressed by the Stack Pointer is read, and the Stack Pointer is decremented by one. The value read is then transferred to the directly addressed byte indicated. No flags are affected. |
| **Example:** | The Stack Pointer originally contains the value 32H, and internal RAM locations 30H through 32H contain the values 20H, 23H, and 01H, respectively. The instruction sequence, |

POP   DPH

POP   DPL

will leave the Stack Pointer equal to the value 30H and the Data Pointer set to 0123H. At this point the instruction,

POP   SP

will leave the Stack Pointer set to 20H. Note that in this special case the Stack Pointer was decremented to 2FH before being loaded with the value popped (20H).

| | |
|---|---|
| **Bytes:** | 2 |
| **Cycles:** | 2 |
| **Encoding:** | `1 1 0 1` `0 0 0 0`    `direct address` |
| **Operation:** | POP<br>(direct) ← ((SP))<br>(SP) ← (SP) − 1 |

---

## PUSH direct

| | |
|---|---|
| **Function:** | Push onto stack |
| **Description:** | The Stack Pointer is incremented by one. The contents of the indicated variable is then copied into the internal RAM location addressed by the Stack Pointer. Otherwise no flags are affected. |
| **Example:** | On entering an interrupt routine the Stack Pointer contains 09H. The Data Pointer holds the value 0123H. The instruction sequence, |

PUSH   DPL

PUSH   DPH

will leave the Stack Pointer set to 0BH and store 23H and 01H in internal RAM locations 0AH and 0BH, respectively.

| | |
|---|---|
| **Bytes:** | 2 |
| **Cycles:** | 2 |
| **Encoding:** | `1 1 0 0` `0 0 0 0`    `direct address` |
| **Operation:** | PUSH<br>(SP) ← (SP) + 1<br>((SP)) ← (direct) |

## RET

| | |
|---|---|
| **Function:** | Return from subroutine |
| **Description:** | RET pops the high- and low-order bytes of the PC successively from the stack, decrementing the Stack Pointer by two. Program execution continues at the resulting address, generally the instruction immediately following an ACALL or LCALL. No flags are affected. |
| **Example:** | The Stack Pointer originally contains the value 0BH. Internal RAM locations 0AH and 0BH contain the values 23H and 01H, respectively. The instruction, |

RET

will leave the Stack Pointer equal to the value 09H. Program execution will continue at location 0123H.

| | |
|---|---|
| **Bytes:** | 1 |
| **Cycles:** | 2 |
| **Encoding:** | 0 0 1 0 \| 0 0 1 0 |
| **Operation:** | RET |

$$(PC_{15-8}) \leftarrow ((SP))$$
$$(SP) \leftarrow (SP) - 1$$
$$(PC_{7-0}) \leftarrow ((SP))$$
$$(SP) \leftarrow (SP) - 1$$

## RETI

| | |
|---|---|
| **Function:** | Return from interrupt |
| **Description:** | RETI pops the high- and low-order bytes of the PC successively from the stack, and restores the interrupt logic to accept additional interrupts at the same priority level as the one just processed. The Stack Pointer is left decremented by two. No other registers are affected; the PSW is *not* automatically restored to its pre-interrupt status. Program execution continues at the resulting address, which is generally the instruction immediately after the point at which the interrupt request was detected. If a lower- or same-level interrupt had been pending when the RETI instruction is executed, that one instruction will be executed before the pending interrupt is processed. |
| **Example:** | The Stack Pointer originally contains the value 0BH. An interrupt was detected during the instruction ending at location 0122H. Internal RAM locations 0AH and 0BH contain the values 23H and 01H, respectively. The instruction, |

RETI

will leave the Stack Pointer equal to 09H and return program execution to location 0123H.

| | |
|---|---|
| **Bytes:** | 1 |
| **Cycles:** | 2 |
| **Encoding:** | 0 0 1 1 \| 0 0 1 0 |
| **Operation:** | RETI |

$$(PC_{15-8}) \leftarrow ((SP))$$
$$(SP) \leftarrow (SP) - 1$$
$$(PC_{7-0}) \leftarrow ((SP))$$
$$(SP) \leftarrow (SP) - 1$$

250

## RL   A

| | |
|---|---|
| **Function:** | Rotate Accumulator Left |
| **Description:** | The eight bits in the Accumulator are rotated one bit to the left. Bit 7 is rotated into the bit 0 position. No flags are affected. |
| **Example:** | The Accumulator holds the value 0C5H (11000101B). The instruction, |

RL   A

leaves the Accumulator holding the value 8BH (10001011B) with the carry unaffected.

| | |
|---|---|
| **Bytes:** | 1 |
| **Cycles:** | 1 |
| **Encoding:** | 0 0 1 0    0 0 1 1 |
| **Operation:** | RL |

$$(A_n + 1) \leftarrow (An) \quad n = 0 - 6$$
$$(A0) \leftarrow (A7)$$

## RLC   A

| | |
|---|---|
| **Function:** | Rotate Accumulator Left through the Carry flag |
| **Description:** | The eight bits in the Accumulator and the carry flag are together rotated one bit to the left. Bit 7 moves into the carry flag; the original state of the carry flag moves into the bit 0 position. No other flags are affected. |
| **Example:** | The Accumulator holds the value 0C5H (11000101B), and the carry is zero. The instruction, |

RLC   A

leaves the Accumulator holding the value 8BH (10001010B) with the carry set.

| | |
|---|---|
| **Bytes:** | 1 |
| **Cycles:** | 1 |
| **Encoding:** | 0 0 1 1    0 0 1 1 |
| **Operation:** | RLC |

$$(An + 1) \leftarrow (An) \quad n = 0 - 6$$
$$(A0) \leftarrow (C)$$
$$(C) \leftarrow (A7)$$

## RR  A

| | |
|---|---|
| **Function:** | Rotate Accumulator Right |
| **Description:** | The eight bits in the Accumulator are rotated one bit to the right. Bit 0 is rotated into the bit 7 position. No flags are affected. |
| **Example:** | The Accumulator holds the value 0C5H (11000101B). The instruction, |

RR   A

leaves the Accumulator holding the value 0E2H (11100010B) with the carry unaffected.

| | |
|---|---|
| **Bytes:** | 1 |
| **Cycles:** | 1 |
| **Encoding:** | 0 0 0 0 \| 0 0 1 1 |
| **Operation:** | RR |

$$(A_n) \leftarrow (A_n + 1) \quad n = 0 - 6$$
$$(A7) \leftarrow (A0)$$

## RRC  A

| | |
|---|---|
| **Function:** | Rotate Accumulator Right through Carry flag |
| **Description:** | The eight bits in the Accumulator and the carry flag are together rotated one bit to the right. Bit 0 moves into the carry flag; the original value of the carry flag moves into the bit 7 position. No other flags are affected. |
| **Example:** | The Accumulator holds the value 0C5H (11000101B), the carry is zero. The instruction, |

RRC   A

leaves the Accumulator holding the value 62 (01100010B) with the carry set.

| | |
|---|---|
| **Bytes:** | 1 |
| **Cycles:** | 1 |
| **Encoding:** | 0 0 0 1 \| 0 0 1 1 |
| **Operation:** | RRC |

$$(A_n) \leftarrow (A_n + 1) \quad n = 0 - 6$$
$$(A7) \leftarrow (C)$$
$$(C) \leftarrow (A0)$$

---

**SETB**  &lt;bit&gt;

---

| | |
|---|---|
| **Function:** | Set Bit |
| **Description:** | SETB sets the indicated bit to one. SETB can operate on the carry flag or any directly addressable bit. No other flags are affected. |
| **Example:** | The carry flag is cleared. Output Port 1 has been written with the value 34H (00110100B). The instructions, |

SETB   C

SETB   P1.0

will leave the carry flag set to 1 and change the data output on Port 1 to 35H (00110101B).

**SETB   C**

| | |
|---|---|
| **Bytes:** | 1 |
| **Cycles:** | 1 |
| **Encoding:** | 1 1 0 1 \| 0 0 1 1 |
| **Operation:** | SETB<br>(C) ← 1 |

**SETB   bit**

| | |
|---|---|
| **Bytes:** | 2 |
| **Cycles:** | 1 |
| **Encoding:** | 1 1 0 1 \| 0 0 1 0 \| bit address |
| **Operation:** | SETB<br>(bit) ← 1 |

---

**SJMP   rel**

---

|  |  |
|---|---|
| **Function:** | Short Jump |
| **Description:** | Program control branches unconditionally to the address indicated. The branch destination is computed by adding the signed displacement in the second instruction byte to the PC, after incrementing the PC twice. Therefore, the range of destinations allowed is from 128 bytes preceding this instruction to 127 bytes following it. |
| **Example:** | The label "RELADR" is assigned to an instruction at program memory location 0123H. The instruction, |

SJMP   RELADR

will assemble into location 0100H. After the instruction is executed, the PC will contain the value 0123H.

(*Note:* Under the above conditions the instruction following SJMP will be at 102H. Therefore, the displacement byte of the instruction will be the relative offset (0123H-0102H) = 21H. Put another way, an SJMP with a displacement of 0FEH would be a one-instruction infinite loop.)

|  |  |
|---|---|
| **Bytes:** | 2 |
| **Cycles:** | 2 |
| **Encoding:** | 1 0 0 0  │  0 0 0 0        rel. address |
| **Operation:** | SJMP<br>(PC) ← (PC) + 2<br>(PC) ← (PC) + rel |

## SUBB   A, <src-byte>

| | |
|---|---|
| **Function:** | Subtract with borrow |
| **Description:** | SUBB subtracts the indicated variable and the carry flag together from the Accumulator, leaving the result in the Accumulator. SUBB sets the carry (borrow) flag if a borrow is needed for bit 7, and clears C otherwise. (If C was set *before* executing a SUBB instruction, this indicates that a borrow was needed for the previous step in a multiple precision subtraction, so the carry is subtracted from the Accumulator along with the source operand.) AC is set if a borrow is needed for bit 3, and cleared otherwise. OV is set if a borrow is needed into bit 6, but not into bit 7, or into bit 7, but not bit 6. |

When subtracting signed integers OV indicates a negative number produced when a negative value is subtracted from a positive value, or a positive result when a positive number is subtracted from a negative number.

The source operand allows four addressing modes: register, direct, register-indirect, or immediate.

| | |
|---|---|
| **Example:** | The Accumulator holds 0C9H (11001001B), register 2 holds 54H (01010100B), and the carry flag is set. The instruction, |

SUBB   A,R2

will leave the value 74H (01110100B) in the accumulator, with the carry flag and AC cleared but OV set.

Notice that 0C9H minus 54H is 75H. The difference between this and the above result is due to the carry (borrow) flag being set before the operation. If the state of the carry is not known before starting a single or multiple-precision subtraction, it should be explicitly cleared by a CLR C instruction.

## SUBB   A,Rn

| | |
|---|---|
| **Bytes:** | 1 |
| **Cycles:** | 1 |
| **Encoding:** | 1 0 0 1 \| 1 r r r |
| **Operation:** | SUBB<br>(A) ← (A) − (C) − (Rn) |

**SUBB   A,direct**

| | |
|---|---|
| **Bytes:** | 2 |
| **Cycles:** | 1 |

**Encoding:**

| 1 0 0 1 | 0 1 0 1 | | direct address |
|---|---|---|---|

**Operation:**  SUBB
$(A) \leftarrow (A) - (C) - (direct)$

**SUBB   A,@Ri**

| | |
|---|---|
| **Bytes:** | 1 |
| **Cycles:** | 1 |

**Encoding:**

| 1 0 0 1 | 0 1 1 i |
|---|---|

**Operation:**  SUBB
$(A) \leftarrow (A) - (C) - ((Ri))$

**SUBB   A,#data**

| | |
|---|---|
| **Bytes:** | 2 |
| **Cycles:** | 1 |

**Encoding:**

| 1 0 0 1 | 0 1 0 0 | | immediate data |
|---|---|---|---|

**Operation:**  SUBB
$(A) \leftarrow (A) - (C) - \#data$

---

**SWAP   A**

| | |
|---|---|
| **Function:** | Swap nibbles within the Accumulator |
| **Description:** | SWAP A interchanges the low- and high-order nibbles (four-bit fields) of the Accumulator (bits 3-0 and bits 7-4). The operation can also be thought of as a four-bit rotate instruction. No flags are affected. |
| **Example:** | The Accumulator holds the value 0C5H (11000101B). The instruction, |

SWAP   A

leaves the Accumulator holding the value 5CH (01011100B).

| | |
|---|---|
| **Bytes:** | 1 |
| **Cycles:** | 1 |

**Encoding:**

| 1 1 0 0 | 0 1 0 0 |
|---|---|

**Operation:**  SWAP
$(A_{3-0}) \rightleftarrows (A_{7-4})$

## XCH  A,<byte>

| | |
|---|---|
| **Function:** | Exchange Accumulator with byte variable |
| **Description:** | XCH loads the Accumulator with the contents of the indicated variable, at the same time writing the original Accumulator contents to the indicated variable. The source/destination operand can use register, direct, or register-indirect addressing. |
| **Example:** | R0 contains the address 20H. The Accumulator holds the value 3FH (00111111B). Internal RAM location 20H holds the value 75H (01110101B). The instruction, |

XCH  A,@R0

will leave RAM location 20H holding the values 3FH (00111111B) and 75H (01110101B) in the accumulator.

## XCH  A,Rn

| | |
|---|---|
| **Bytes:** | 1 |
| **Cycles:** | 1 |
| **Encoding:** | `1 1 0 0` `1 r r r` |
| **Operation:** | XCH<br>(A) $\rightleftarrows$ (Rn) |

## XCH  A,direct

| | |
|---|---|
| **Bytes:** | 2 |
| **Cycles:** | 1 |
| **Encoding:** | `1 1 0 0` `0 1 0 1`   `direct address` |
| **Operation:** | XCH<br>(A) $\rightleftarrows$ (direct) |

## XCH  A,@Ri

| | |
|---|---|
| **Bytes:** | 1 |
| **Cycles:** | 1 |
| **Encoding:** | `1 1 0 0` `0 1 1 i` |
| **Operation:** | XCH<br>(A) $\rightleftarrows$ ((Ri)) |

## XCHD   A,@Ri

| | |
|---|---|
| **Function:** | Exchange Digit |
| **Description:** | XCHD exchanges the low-order nibble of the Accumulator (bits 3-0), generally representing a hexadecimal or BCD digit, with that of the internal RAM location indirectly addressed by the specified register. The high-order nibbles (bits 7-4) of each register are not affected. No flags are affected. |
| **Example:** | R0 contains the address 20H. The Accumulator holds the value 36H (00110110B). Internal RAM location 20H holds the value 75H (01110101B). The instruction, |

XCHD   A,@R0

will leave RAM location 20H holding the value 76H (01110110B) and 35H (00110101B) in the Accumulator.

| | |
|---|---|
| **Bytes:** | 1 |
| **Cycles:** | 1 |

**Encoding:**

| 1 1 0 1 | 0 1 1 i |
|---|---|

**Operation:**   XCHD
$(A_{3-0}) \rightleftarrows ((Ri_{3-0}))$

## XRL   <dest-byte>,<src-byte>

| | |
|---|---|
| **Function:** | Logical Exclusive-OR for byte variables |
| **Description:** | XRL performs the bitwise logical Exclusive-OR operation between the indicated variables, storing the results in the destination. No flags are affected. |

The two operands allow six addressing mode combinations. When the destination is the Accumulator, the source can use register, direct, register-indirect, or immediate addressing; when the destination is a direct address, the source can be the Accumulator or immediate data.

(*Note:* When this instruction is used to modify an output port, the value used as the original port data will be read from the output data latch, *not* the input pins.)

| | |
|---|---|
| **Example:** | If the Accumulator holds 0C3H (11000011B) and register 0 holds 0AAH (10101010B) then the instruction, |

XRL   A,R0

will leave the Accumulator holding the value 69H (01101001B).

When the destination is a directly addressed byte, this instruction can complement combinations of bits in any RAM location or hardware register. The pattern of bits to be complemented is then determined by a mask byte, either a constant contained in the instruction or a variable computed in the Accumulator at run-time. The instruction,

XRL   P1,#00110001B

will complement bits 5, 4, and 0 of output Port 1.

258

**XRL   A,Rn**

       **Bytes:**  1

       **Cycles:**  1

     **Encoding:**

| 0 1 1 0 | 1 r r r |
|---------|---------|

     **Operation:**  XRL
$(A) \leftarrow (A) \veebar (Rn)$

**XRL   A,direct**

       **Bytes:**  2

       **Cycles:**  1

     **Encoding:**

| 0 1 1 0 | 0 1 0 1 |    direct address |
|---------|---------|------------------|

     **Operation:**  XRL
$(A) \leftarrow (A) \veebar (direct)$

**XRL   A,@Ri**

       **Bytes:**  1

       **Cycles:**  1

     **Encoding:**

| 0 1 1 0 | 0 1 1 i |
|---------|---------|

     **Operation:**  XRL
$(A) \leftarrow (A) \veebar ((Ri))$

**XRL   A,#data**

       **Bytes:**  2

       **Cycles:**  1

     **Encoding:**

| 0 1 1 0 | 0 1 0 0 |    immediate data |
|---------|---------|-------------------|

     **Operation:**  XRL
$(A) \leftarrow (A) \veebar \#data$

**XRL   direct,A**

       **Bytes:**  2

       **Cycles:**  1

     **Encoding:**

| 0 1 1 0 | 0 0 1 0 |    direct address |
|---------|---------|------------------|

     **Operation:**  XRL
$(direct) \leftarrow (direct) \veebar (A)$

**XRL    direct,#data**

>           **Bytes:**    3

>          **Cycles:**    2

>        **Encoding:**    | 0 1 1 0 | 0 0 1 1 |    | direct address |    | immediate data |

>       **Operation:**    XRL
>                         (direct) ← (direct) ∀ #data

# ASCII and EBCDIC Tables

## ASCII-77 CODE—ODD PARITY

| | \ | | Binary Code | | | | | | Hex | | Binary Code | | | | | | | | Hex |
|---|---|---|---|---|---|---|---|---|---|---|---|---|---|---|---|---|---|---|---|
| Bit: | 7 | 6 | 5 | 4 | 3 | 2 | 1 | 0 | Hex | Bit: | 7 | 6 | 5 | 4 | 3 | 2 | 1 | 0 | Hex |
| NUL | 1 | 0 | 0 | 0 | 0 | 0 | 0 | 0 | 00 | @ | 0 | 1 | 0 | 0 | 0 | 0 | 0 | 0 | 40 |
| SOH | 0 | 0 | 0 | 0 | 0 | 0 | 0 | 1 | 01 | A | 1 | 1 | 0 | 0 | 0 | 0 | 0 | 1 | 41 |
| STX | 0 | 0 | 0 | 0 | 0 | 0 | 1 | 0 | 02 | B | 1 | 1 | 0 | 0 | 0 | 0 | 1 | 0 | 42 |
| ETX | 1 | 0 | 0 | 0 | 0 | 0 | 1 | 1 | 03 | C | 0 | 1 | 0 | 0 | 0 | 0 | 1 | 1 | 43 |
| EOT | 0 | 0 | 0 | 0 | 0 | 1 | 0 | 0 | 04 | D | 1 | 1 | 0 | 0 | 0 | 1 | 0 | 0 | 44 |
| ENQ | 1 | 0 | 0 | 0 | 0 | 1 | 0 | 1 | 05 | E | 0 | 1 | 0 | 0 | 0 | 1 | 0 | 1 | 45 |
| ACK | 1 | 0 | 0 | 0 | 0 | 1 | 1 | 0 | 06 | F | 0 | 1 | 0 | 0 | 0 | 1 | 1 | 0 | 46 |
| BEL | 0 | 0 | 0 | 0 | 0 | 1 | 1 | 1 | 07 | G | 1 | 1 | 0 | 0 | 0 | 1 | 1 | 1 | 47 |
| BS | 0 | 0 | 0 | 0 | 1 | 0 | 0 | 0 | 08 | H | 1 | 1 | 0 | 0 | 1 | 0 | 0 | 0 | 48 |
| HT | 1 | 0 | 0 | 0 | 1 | 0 | 0 | 1 | 09 | I | 0 | 1 | 0 | 0 | 1 | 0 | 0 | 1 | 49 |
| NL | 1 | 0 | 0 | 0 | 1 | 0 | 1 | 0 | 0A | J | 0 | 1 | 0 | 0 | 1 | 0 | 1 | 0 | 4A |
| VT | 0 | 0 | 0 | 0 | 1 | 0 | 1 | 1 | 0B | K | 1 | 1 | 0 | 0 | 1 | 0 | 1 | 1 | 4B |
| FF | 1 | 0 | 0 | 0 | 1 | 1 | 0 | 0 | 0C | L | 0 | 1 | 0 | 0 | 1 | 1 | 0 | 0 | 4C |
| CR | 0 | 0 | 0 | 0 | 1 | 1 | 0 | 1 | 0D | M | 1 | 1 | 0 | 0 | 1 | 1 | 0 | 1 | 4D |
| SO | 0 | 0 | 0 | 0 | 1 | 1 | 1 | 0 | 0E | N | 1 | 1 | 0 | 0 | 1 | 1 | 1 | 0 | 4E |
| SI | 1 | 0 | 0 | 0 | 1 | 1 | 1 | 1 | 0F | O | 0 | 1 | 0 | 0 | 1 | 1 | 1 | 1 | 4F |
| DLE | 0 | 0 | 0 | 1 | 0 | 0 | 0 | 0 | 10 | P | 1 | 1 | 0 | 1 | 0 | 0 | 0 | 0 | 50 |
| DC1 | 0 | 0 | 0 | 1 | 0 | 0 | 0 | 1 | 11 | Q | 0 | 1 | 0 | 1 | 0 | 0 | 0 | 1 | 51 |
| DC2 | 1 | 0 | 0 | 1 | 0 | 0 | 1 | 0 | 12 | R | 0 | 1 | 0 | 1 | 0 | 0 | 1 | 0 | 52 |
| DC3 | 0 | 0 | 0 | 1 | 0 | 0 | 1 | 1 | 13 | S | 1 | 1 | 0 | 1 | 0 | 0 | 1 | 1 | 53 |
| DC4 | 1 | 0 | 0 | 1 | 0 | 1 | 0 | 0 | 14 | T | 0 | 1 | 0 | 1 | 0 | 1 | 0 | 0 | 54 |
| NAK | 0 | 0 | 0 | 1 | 0 | 1 | 0 | 1 | 15 | U | 1 | 1 | 0 | 1 | 0 | 1 | 0 | 1 | 55 |
| SYN | 0 | 0 | 0 | 1 | 0 | 1 | 1 | 0 | 16 | V | 1 | 1 | 0 | 1 | 0 | 1 | 1 | 0 | 56 |
| ETB | 1 | 0 | 0 | 1 | 0 | 1 | 1 | 1 | 17 | W | 0 | 1 | 0 | 1 | 0 | 1 | 1 | 1 | 57 |
| CAN | 1 | 0 | 0 | 1 | 1 | 0 | 0 | 0 | 18 | X | 0 | 1 | 0 | 1 | 1 | 0 | 0 | 0 | 58 |
| EM | 0 | 0 | 0 | 1 | 1 | 0 | 0 | 1 | 19 | Y | 1 | 1 | 0 | 1 | 1 | 0 | 0 | 1 | 59 |
| SUB | 0 | 0 | 0 | 1 | 1 | 0 | 1 | 0 | 1A | Z | 1 | 1 | 0 | 1 | 1 | 0 | 1 | 0 | 5A |
| ESC | 1 | 0 | 0 | 1 | 1 | 0 | 1 | 1 | 1B | [ | 0 | 1 | 0 | 1 | 1 | 0 | 1 | 1 | 5B |
| FS | 0 | 0 | 0 | 1 | 1 | 1 | 0 | 0 | 1C | \ | 1 | 1 | 0 | 1 | 1 | 1 | 0 | 0 | 5C |
| GS | 1 | 0 | 0 | 1 | 1 | 1 | 0 | 1 | 1D | ] | 0 | 1 | 0 | 1 | 1 | 1 | 0 | 1 | 5D |
| RS | 1 | 0 | 0 | 1 | 1 | 1 | 1 | 0 | 1E | ∧ | 0 | 1 | 0 | 1 | 1 | 1 | 1 | 0 | 5E |
| US | 0 | 0 | 0 | 1 | 1 | 1 | 1 | 1 | 1F | — | 1 | 1 | 0 | 1 | 1 | 1 | 1 | 1 | 5F |
| SP | 0 | 0 | 1 | 0 | 0 | 0 | 0 | 0 | 20 | ` | 1 | 1 | 1 | 0 | 0 | 0 | 0 | 0 | 60 |
| ! | 1 | 0 | 1 | 0 | 0 | 0 | 0 | 1 | 21 | a | 0 | 1 | 1 | 0 | 0 | 0 | 0 | 1 | 61 |
| " | 1 | 0 | 1 | 0 | 0 | 0 | 1 | 0 | 22 | b | 0 | 1 | 1 | 0 | 0 | 0 | 1 | 0 | 62 |
| # | 0 | 0 | 1 | 0 | 0 | 0 | 1 | 1 | 23 | c | 1 | 1 | 1 | 0 | 0 | 0 | 1 | 1 | 63 |
| $ | 1 | 0 | 1 | 0 | 0 | 1 | 0 | 0 | 24 | d | 0 | 1 | 1 | 0 | 0 | 1 | 0 | 0 | 64 |
| % | 0 | 0 | 1 | 0 | 0 | 1 | 0 | 1 | 25 | e | 1 | 1 | 1 | 0 | 0 | 1 | 0 | 1 | 65 |
| & | 0 | 0 | 1 | 0 | 0 | 1 | 1 | 0 | 26 | f | 1 | 1 | 1 | 0 | 0 | 1 | 1 | 0 | 66 |
| ' | 1 | 0 | 1 | 0 | 0 | 1 | 1 | 1 | 27 | g | 0 | 1 | 1 | 0 | 0 | 1 | 1 | 1 | 67 |
| ( | 1 | 0 | 1 | 0 | 1 | 0 | 0 | 0 | 28 | h | 0 | 1 | 1 | 0 | 1 | 0 | 0 | 0 | 68 |
| ) | 0 | 0 | 1 | 0 | 1 | 0 | 0 | 1 | 29 | i | 1 | 1 | 1 | 0 | 1 | 0 | 0 | 1 | 69 |
| * | 0 | 0 | 1 | 0 | 1 | 0 | 1 | 0 | 2A | j | 1 | 1 | 1 | 0 | 1 | 0 | 1 | 0 | 6A |
| + | 1 | 0 | 1 | 0 | 1 | 0 | 1 | 1 | 2B | k | 0 | 1 | 1 | 0 | 1 | 0 | 1 | 1 | 6B |
| , | 0 | 0 | 1 | 0 | 1 | 1 | 0 | 0 | 2C | l | 1 | 1 | 1 | 0 | 1 | 1 | 0 | 0 | 6C |
| - | 1 | 0 | 1 | 0 | 1 | 1 | 0 | 1 | 2D | m | 0 | 1 | 1 | 0 | 1 | 1 | 0 | 1 | 6D |

| | Binary Code | | | | | | | | Hex | | Binary Code | | | | | | | | Hex |
|---|---|---|---|---|---|---|---|---|---|---|---|---|---|---|---|---|---|---|---|---|
| Bit: | 7 | 6 | 5 | 4 | 3 | 2 | 1 | 0 | | Bit: | 7 | 6 | 5 | 4 | 3 | 2 | 1 | 0 | |
| . | 1 | 0 | 1 | 0 | 1 | 1 | 1 | 0 | 2E | n | 0 | 1 | 1 | 0 | 1 | 1 | 1 | 0 | 6E |
| / | 0 | 0 | 1 | 0 | 1 | 1 | 1 | 1 | 2F | o | 1 | 1 | 1 | 0 | 1 | 1 | 1 | 1 | 6F |
| 0 | 1 | 0 | 1 | 1 | 0 | 0 | 0 | 0 | 30 | p | 0 | 1 | 1 | 1 | 0 | 0 | 0 | 0 | 70 |
| 1 | 0 | 0 | 1 | 1 | 0 | 0 | 0 | 1 | 31 | q | 1 | 1 | 1 | 1 | 0 | 0 | 0 | 1 | 71 |
| 2 | 0 | 0 | 1 | 1 | 0 | 0 | 1 | 0 | 32 | r | 1 | 1 | 1 | 1 | 0 | 0 | 1 | 0 | 72 |
| 3 | 1 | 0 | 1 | 1 | 0 | 0 | 1 | 1 | 33 | s | 0 | 1 | 1 | 1 | 0 | 0 | 1 | 1 | 73 |
| 4 | 0 | 0 | 1 | 1 | 0 | 1 | 0 | 0 | 34 | t | 1 | 1 | 1 | 1 | 0 | 1 | 0 | 0 | 74 |
| 5 | 1 | 0 | 1 | 1 | 0 | 1 | 0 | 1 | 35 | u | 0 | 1 | 1 | 1 | 0 | 1 | 0 | 1 | 75 |
| 6 | 1 | 0 | 1 | 1 | 0 | 1 | 1 | 0 | 36 | v | 0 | 1 | 1 | 1 | 0 | 1 | 1 | 0 | 76 |
| 7 | 0 | 0 | 1 | 1 | 0 | 1 | 1 | 1 | 37 | w | 1 | 1 | 1 | 1 | 0 | 1 | 1 | 1 | 77 |
| 8 | 0 | 0 | 1 | 1 | 1 | 0 | 0 | 0 | 38 | x | 1 | 1 | 1 | 1 | 1 | 0 | 0 | 0 | 78 |
| 9 | 1 | 0 | 1 | 1 | 1 | 0 | 0 | 1 | 39 | y | 0 | 1 | 1 | 1 | 1 | 0 | 0 | 1 | 79 |
| : | 1 | 0 | 1 | 1 | 1 | 0 | 1 | 0 | 3A | z | 0 | 1 | 1 | 1 | 1 | 0 | 1 | 0 | 7A |
| ; | 0 | 0 | 1 | 1 | 1 | 0 | 1 | 1 | 3B | { | 1 | 1 | 1 | 1 | 1 | 0 | 1 | 1 | 7B |
| < | 1 | 0 | 1 | 1 | 1 | 1 | 0 | 0 | 3C | ¦ | 0 | 1 | 1 | 1 | 1 | 1 | 0 | 0 | 7C |
| = | 0 | 0 | 1 | 1 | 1 | 1 | 0 | 1 | 3D | } | 1 | 1 | 1 | 1 | 1 | 1 | 0 | 1 | 7D |
| > | 0 | 0 | 1 | 1 | 1 | 1 | 1 | 0 | 3E | ~ | 1 | 1 | 1 | 1 | 1 | 1 | 1 | 0 | 7E |
| ? | 1 | 0 | 1 | 1 | 1 | 1 | 1 | 1 | 3F | DEL | 0 | 1 | 1 | 1 | 1 | 1 | 1 | 1 | 7F |

| | | | | |
|---|---|---|---|---|
| NUL | = null | | DC1 | = device control 1 |
| SOH | = start of heading | | DC2 | = device control 2 |
| STX | = start of text | | DC3 | = device control 3 |
| ETX | = end of text | | DC4 | = device control 4 |
| EOT | = end of transmission | | NAK | = negative acknowledge |
| ENQ | = enquiry | | SYN | = synchronous |
| ACK | = acknowledge | | ETB | = end of transmission block |
| BEL | = bell | | CAN | = cancel |
| BS | = back space | | SUB | = substitute |
| HT | = horizontal tab | | ESC | = escape |
| NL | = new line | | FS | = field separator |
| VT | = vertical tab | | GS | = group separator |
| FF | = form feed | | RS | = record separator |
| CR | = carriage return | | US | = unit separator |
| SO | = shift-out | | SP | = space |
| SI | = shift-in | | DEL | = delete |
| DLE | = data link escape | | | |

Note: Standard ASCII is a 7-bit code. The parity bit (bit 7) is often not used. The Hex numbers given correspond to standard ASCII.

# EBCDIC CODE

| | Binary Code | | | | | | | | Hex | | Binary Code | | | | | | | | Hex |
|---|---|---|---|---|---|---|---|---|---|---|---|---|---|---|---|---|---|---|---|---|
| Bit: | 0 | 1 | 2 | 3 | 4 | 5 | 6 | 7 | | Bit: | 0 | 1 | 2 | 3 | 4 | 5 | 6 | 7 | |
| NUL | 0 | 0 | 0 | 0 | 0 | 0 | 0 | 0 | 00 | | 1 | 0 | 0 | 0 | 0 | 0 | 0 | 0 | 80 |
| SOH | 0 | 0 | 0 | 0 | 0 | 0 | 0 | 1 | 01 | a | 1 | 0 | 0 | 0 | 0 | 0 | 0 | 1 | 81 |
| STX | 0 | 0 | 0 | 0 | 0 | 0 | 1 | 0 | 02 | b | 1 | 0 | 0 | 0 | 0 | 0 | 1 | 0 | 82 |
| ETX | 0 | 0 | 0 | 0 | 0 | 0 | 1 | 1 | 03 | c | 1 | 0 | 0 | 0 | 0 | 0 | 1 | 1 | 83 |
| | 0 | 0 | 0 | 0 | 0 | 1 | 0 | 0 | 04 | d | 1 | 0 | 0 | 0 | 0 | 1 | 0 | 0 | 84 |
| PT | 0 | 0 | 0 | 0 | 0 | 1 | 0 | 1 | 05 | e | 1 | 0 | 0 | 0 | 0 | 1 | 0 | 1 | 85 |
| | 0 | 0 | 0 | 0 | 0 | 1 | 1 | 0 | 06 | f | 1 | 0 | 0 | 0 | 0 | 1 | 1 | 0 | 86 |
| | 0 | 0 | 0 | 0 | 0 | 1 | 1 | 1 | 07 | g | 1 | 0 | 0 | 0 | 0 | 1 | 1 | 1 | 87 |
| | 0 | 0 | 0 | 0 | 1 | 0 | 0 | 0 | 08 | h | 1 | 0 | 0 | 0 | 1 | 0 | 0 | 0 | 88 |
| | 0 | 0 | 0 | 0 | 1 | 0 | 0 | 1 | 09 | i | 1 | 0 | 0 | 0 | 1 | 0 | 0 | 1 | 89 |
| | 0 | 0 | 0 | 0 | 1 | 0 | 1 | 0 | 0A | | 1 | 0 | 0 | 0 | 1 | 0 | 1 | 0 | 8A |
| | 0 | 0 | 0 | 0 | 1 | 0 | 1 | 1 | 0B | | 1 | 0 | 0 | 0 | 1 | 0 | 1 | 1 | 8B |
| FF | 0 | 0 | 0 | 0 | 1 | 1 | 0 | 0 | 0C | | 1 | 0 | 0 | 0 | 1 | 1 | 0 | 0 | 8C |
| | 0 | 0 | 0 | 0 | 1 | 1 | 0 | 1 | 0D | | 1 | 0 | 0 | 0 | 1 | 1 | 0 | 1 | 8D |
| | 0 | 0 | 0 | 0 | 1 | 1 | 1 | 0 | 0E | | 1 | 0 | 0 | 0 | 1 | 1 | 1 | 0 | 8E |
| | 0 | 0 | 0 | 0 | 1 | 1 | 1 | 1 | 0F | | 1 | 0 | 0 | 0 | 1 | 1 | 1 | 1 | 8F |
| DLE | 0 | 0 | 0 | 1 | 0 | 0 | 0 | 0 | 10 | | 1 | 0 | 0 | 1 | 0 | 0 | 0 | 0 | 90 |
| SBA | 0 | 0 | 0 | 1 | 0 | 0 | 0 | 1 | 11 | j | 1 | 0 | 0 | 1 | 0 | 0 | 0 | 1 | 91 |
| EUA | 0 | 0 | 0 | 1 | 0 | 0 | 1 | 0 | 12 | k | 1 | 0 | 0 | 1 | 0 | 0 | 1 | 0 | 92 |
| IC | 0 | 0 | 0 | 1 | 0 | 0 | 1 | 1 | 13 | l | 1 | 0 | 0 | 1 | 0 | 0 | 1 | 1 | 93 |
| | 0 | 0 | 0 | 1 | 0 | 1 | 0 | 0 | 14 | m | 1 | 0 | 0 | 1 | 0 | 1 | 0 | 0 | 94 |
| NL | 0 | 0 | 0 | 1 | 0 | 1 | 0 | 1 | 15 | n | 1 | 0 | 0 | 1 | 0 | 1 | 0 | 1 | 95 |
| | 0 | 0 | 0 | 1 | 0 | 1 | 1 | 0 | 16 | o | 1 | 0 | 0 | 1 | 0 | 1 | 1 | 0 | 96 |
| | 0 | 0 | 0 | 1 | 0 | 1 | 1 | 1 | 17 | p | 1 | 0 | 0 | 1 | 0 | 1 | 1 | 1 | 97 |
| | 0 | 0 | 0 | 1 | 1 | 0 | 0 | 0 | 18 | q | 1 | 0 | 0 | 1 | 1 | 0 | 0 | 0 | 98 |
| EM | 0 | 0 | 0 | 1 | 1 | 0 | 0 | 1 | 19 | r | 1 | 0 | 0 | 1 | 1 | 0 | 0 | 1 | 99 |
| | 0 | 0 | 0 | 1 | 1 | 0 | 1 | 0 | 1A | | 1 | 0 | 0 | 1 | 1 | 0 | 1 | 0 | 9A |
| | 0 | 0 | 0 | 1 | 1 | 0 | 1 | 1 | 1B | | 1 | 0 | 0 | 1 | 1 | 0 | 1 | 1 | 9B |
| DUP | 0 | 0 | 0 | 1 | 1 | 1 | 0 | 0 | 1C | | 1 | 0 | 0 | 1 | 1 | 1 | 0 | 0 | 9C |
| SF | 0 | 0 | 0 | 1 | 1 | 1 | 0 | 1 | 1D | | 1 | 0 | 0 | 1 | 1 | 1 | 0 | 1 | 9D |
| FM | 0 | 0 | 0 | 1 | 1 | 1 | 1 | 0 | 1E | | 1 | 0 | 0 | 1 | 1 | 1 | 1 | 0 | 9E |
| ITB | 0 | 0 | 0 | 1 | 1 | 1 | 1 | 1 | 1F | | 1 | 0 | 0 | 1 | 1 | 1 | 1 | 1 | 9F |
| | 0 | 0 | 1 | 0 | 0 | 0 | 0 | 0 | 20 | | 1 | 0 | 1 | 0 | 0 | 0 | 0 | 0 | A0 |
| | 0 | 0 | 1 | 0 | 0 | 0 | 0 | 1 | 21 | ~ | 1 | 0 | 1 | 0 | 0 | 0 | 0 | 1 | A1 |
| | 0 | 0 | 1 | 0 | 0 | 0 | 1 | 0 | 22 | s | 1 | 0 | 1 | 0 | 0 | 0 | 1 | 0 | A2 |
| | 0 | 0 | 1 | 0 | 0 | 0 | 1 | 1 | 23 | t | 1 | 0 | 1 | 0 | 0 | 0 | 1 | 1 | A3 |
| | 0 | 0 | 1 | 0 | 0 | 1 | 0 | 0 | 24 | u | 1 | 0 | 1 | 0 | 0 | 1 | 0 | 0 | A4 |
| | 0 | 0 | 1 | 0 | 0 | 1 | 0 | 1 | 25 | v | 1 | 0 | 1 | 0 | 0 | 1 | 0 | 1 | A5 |
| ETB | 0 | 0 | 1 | 0 | 0 | 1 | 1 | 0 | 26 | w | 1 | 0 | 1 | 0 | 0 | 1 | 1 | 0 | A6 |
| ESC | 0 | 0 | 1 | 0 | 0 | 1 | 1 | 1 | 27 | x | 1 | 0 | 1 | 0 | 0 | 1 | 1 | 1 | A7 |
| | 0 | 0 | 1 | 0 | 1 | 0 | 0 | 0 | 28 | y | 1 | 0 | 1 | 0 | 1 | 0 | 0 | 0 | A8 |
| | 0 | 0 | 1 | 0 | 1 | 0 | 0 | 1 | 29 | z | 1 | 0 | 1 | 0 | 1 | 0 | 0 | 1 | A9 |
| | 0 | 0 | 1 | 0 | 1 | 0 | 1 | 0 | 2A | | 1 | 0 | 1 | 0 | 1 | 0 | 1 | 0 | AA |
| | 0 | 0 | 1 | 0 | 1 | 0 | 1 | 1 | 2B | | 1 | 0 | 1 | 0 | 1 | 0 | 1 | 1 | AB |
| | 0 | 0 | 1 | 0 | 1 | 1 | 0 | 0 | 2C | | 1 | 0 | 1 | 0 | 1 | 1 | 0 | 0 | AC |
| ENQ | 0 | 0 | 1 | 0 | 1 | 1 | 0 | 1 | 2D | | 1 | 0 | 1 | 0 | 1 | 1 | 0 | 1 | AD |
| | 0 | 0 | 1 | 0 | 1 | 1 | 1 | 0 | 2E | | 1 | 0 | 1 | 0 | 1 | 1 | 1 | 0 | AE |
| | 0 | 0 | 1 | 0 | 1 | 1 | 1 | 1 | 2F | | 1 | 0 | 1 | 0 | 1 | 1 | 1 | 1 | AF |
| | 0 | 0 | 1 | 1 | 0 | 0 | 0 | 0 | 30 | | 1 | 0 | 1 | 1 | 0 | 0 | 0 | 0 | B0 |
| | 0 | 0 | 1 | 1 | 0 | 0 | 0 | 1 | 31 | | 1 | 0 | 1 | 1 | 0 | 0 | 0 | 1 | B1 |

# EBCDIC CODE

| | Binary Code | | | | | | | | Hex | | Binary Code | | | | | | | | Hex |
|---|---|---|---|---|---|---|---|---|---|---|---|---|---|---|---|---|---|---|---|
| Bit: | 0 | 1 | 2 | 3 | 4 | 5 | 6 | 7 | | Bit: | 0 | 1 | 2 | 3 | 4 | 5 | 6 | 7 | |
| SYN | 0 | 0 | 1 | 1 | 0 | 0 | 1 | 0 | 32 | | 1 | 0 | 1 | 1 | 0 | 0 | 1 | 0 | B2 |
| | 0 | 0 | 1 | 1 | 0 | 0 | 1 | 1 | 33 | | 1 | 0 | 1 | 1 | 0 | 0 | 1 | 1 | B3 |
| | 0 | 0 | 1 | 1 | 0 | 1 | 0 | 0 | 34 | | 1 | 0 | 1 | 1 | 0 | 1 | 0 | 0 | B4 |
| | 0 | 0 | 1 | 1 | 0 | 1 | 0 | 1 | 35 | | 1 | 0 | 1 | 1 | 0 | 1 | 0 | 1 | B5 |
| | 0 | 0 | 1 | 1 | 0 | 1 | 1 | 0 | 36 | | 1 | 0 | 1 | 1 | 0 | 1 | 1 | 0 | B6 |
| EOT | 0 | 0 | 1 | 1 | 0 | 1 | 1 | 1 | 37 | | 1 | 0 | 1 | 1 | 0 | 1 | 1 | 1 | B7 |
| | 0 | 0 | 1 | 1 | 1 | 0 | 0 | 0 | 38 | | 1 | 0 | 1 | 1 | 1 | 0 | 0 | 0 | B8 |
| | 0 | 0 | 1 | 1 | 1 | 0 | 0 | 1 | 39 | | 1 | 0 | 1 | 1 | 1 | 0 | 0 | 1 | B9 |
| | 0 | 0 | 1 | 1 | 1 | 0 | 1 | 0 | 3A | | 1 | 0 | 1 | 1 | 1 | 0 | 1 | 0 | BA |
| | 0 | 0 | 1 | 1 | 1 | 0 | 1 | 1 | 3B | | 1 | 0 | 1 | 1 | 1 | 0 | 1 | 1 | BB |
| RA | 0 | 0 | 1 | 1 | 1 | 1 | 0 | 0 | 3C | | 1 | 0 | 1 | 1 | 1 | 1 | 0 | 0 | BC |
| NAK | 0 | 0 | 1 | 1 | 1 | 1 | 0 | 1 | 3D | | 1 | 0 | 1 | 1 | 1 | 1 | 0 | 1 | BD |
| | 0 | 0 | 1 | 1 | 1 | 1 | 1 | 0 | 3E | | 1 | 0 | 1 | 1 | 1 | 1 | 1 | 0 | BE |
| SUB | 0 | 0 | 1 | 1 | 1 | 1 | 1 | 1 | 3F | | 1 | 0 | 1 | 1 | 1 | 1 | 1 | 1 | BF |
| SP | 0 | 1 | 0 | 0 | 0 | 0 | 0 | 0 | 40 | { | 1 | 1 | 0 | 0 | 0 | 0 | 0 | 0 | C0 |
| | 0 | 1 | 0 | 0 | 0 | 0 | 0 | 1 | 41 | A | 1 | 1 | 0 | 0 | 0 | 0 | 0 | 1 | C1 |
| | 0 | 1 | 0 | 0 | 0 | 0 | 1 | 0 | 42 | B | 1 | 1 | 0 | 0 | 0 | 0 | 1 | 0 | C2 |
| | 0 | 1 | 0 | 0 | 0 | 0 | 1 | 1 | 43 | C | 1 | 1 | 0 | 0 | 0 | 0 | 1 | 1 | C3 |
| | 0 | 1 | 0 | 0 | 0 | 1 | 0 | 0 | 44 | D | 1 | 1 | 0 | 0 | 0 | 1 | 0 | 0 | C4 |
| | 0 | 1 | 0 | 0 | 0 | 1 | 0 | 1 | 45 | E | 1 | 1 | 0 | 0 | 0 | 1 | 0 | 1 | C5 |
| | 0 | 1 | 0 | 0 | 0 | 1 | 1 | 0 | 46 | F | 1 | 1 | 0 | 0 | 0 | 1 | 1 | 0 | C6 |
| | 0 | 1 | 0 | 0 | 0 | 1 | 1 | 1 | 47 | G | 1 | 1 | 0 | 0 | 0 | 1 | 1 | 1 | C7 |
| | 0 | 1 | 0 | 0 | 1 | 0 | 0 | 0 | 48 | H | 1 | 1 | 0 | 0 | 1 | 0 | 0 | 0 | C8 |
| | 0 | 1 | 0 | 0 | 1 | 0 | 0 | 1 | 49 | I | 1 | 1 | 0 | 0 | 1 | 0 | 0 | 1 | C9 |
| ¢ | 0 | 1 | 0 | 0 | 1 | 0 | 1 | 0 | 4A | | 1 | 1 | 0 | 0 | 1 | 0 | 1 | 0 | CA |
| . | 0 | 1 | 0 | 0 | 1 | 0 | 1 | 1 | 4B | | 1 | 1 | 0 | 0 | 1 | 0 | 1 | 1 | CB |
| < | 0 | 1 | 0 | 0 | 1 | 1 | 0 | 0 | 4C | | 1 | 1 | 0 | 0 | 1 | 1 | 0 | 0 | CC |
| ( | 0 | 1 | 0 | 0 | 1 | 1 | 0 | 1 | 4D | | 1 | 1 | 0 | 0 | 1 | 1 | 0 | 1 | CD |
| + | 0 | 1 | 0 | 0 | 1 | 1 | 1 | 0 | 4E | | 1 | 1 | 0 | 0 | 1 | 1 | 1 | 0 | CE |
| ¦ | 0 | 1 | 0 | 0 | 1 | 1 | 1 | 1 | 4F | | 1 | 1 | 0 | 0 | 1 | 1 | 1 | 1 | CF |
| & | 0 | 1 | 0 | 1 | 0 | 0 | 0 | 0 | 50 | } | 1 | 1 | 0 | 1 | 0 | 0 | 0 | 0 | D0 |
| | 0 | 1 | 0 | 1 | 0 | 0 | 0 | 1 | 51 | J | 1 | 1 | 0 | 1 | 0 | 0 | 0 | 1 | D1 |
| | 0 | 1 | 0 | 1 | 0 | 0 | 1 | 0 | 52 | K | 1 | 1 | 0 | 1 | 0 | 0 | 1 | 0 | D2 |
| | 0 | 1 | 0 | 1 | 0 | 0 | 1 | 1 | 53 | L | 1 | 1 | 0 | 1 | 0 | 0 | 1 | 1 | D3 |
| | 0 | 1 | 0 | 1 | 0 | 1 | 0 | 0 | 54 | M | 1 | 1 | 0 | 1 | 0 | 1 | 0 | 0 | D4 |
| | 0 | 1 | 0 | 1 | 0 | 1 | 0 | 1 | 55 | N | 1 | 1 | 0 | 1 | 0 | 1 | 0 | 1 | D5 |
| | 0 | 1 | 0 | 1 | 0 | 1 | 1 | 0 | 56 | O | 1 | 1 | 0 | 1 | 0 | 1 | 1 | 0 | D6 |
| | 0 | 1 | 0 | 1 | 0 | 1 | 1 | 1 | 57 | P | 1 | 1 | 0 | 1 | 0 | 1 | 1 | 1 | D7 |
| | 0 | 1 | 0 | 1 | 1 | 0 | 0 | 0 | 58 | Q | 1 | 1 | 0 | 1 | 1 | 0 | 0 | 0 | D8 |
| | 0 | 1 | 0 | 1 | 1 | 0 | 0 | 1 | 59 | R | 1 | 1 | 0 | 1 | 1 | 0 | 0 | 1 | D9 |
| ! | 0 | 1 | 0 | 1 | 1 | 0 | 1 | 0 | 5A | | 1 | 1 | 0 | 1 | 1 | 0 | 1 | 0 | DA |
| $ | 0 | 1 | 0 | 1 | 1 | 0 | 1 | 1 | 5B | | 1 | 1 | 0 | 1 | 1 | 0 | 1 | 1 | DB |
| * | 0 | 1 | 0 | 1 | 1 | 1 | 0 | 0 | 5C | | 1 | 1 | 0 | 1 | 1 | 1 | 0 | 0 | DC |
| ) | 0 | 1 | 0 | 1 | 1 | 1 | 0 | 1 | 5D | | 1 | 1 | 0 | 1 | 1 | 1 | 0 | 1 | DD |
| ; | 0 | 1 | 0 | 1 | 1 | 1 | 1 | 0 | 5E | | 1 | 1 | 0 | 1 | 1 | 1 | 1 | 0 | DE |
| ¬ | 0 | 1 | 0 | 1 | 1 | 1 | 1 | 1 | 5F | | 1 | 1 | 0 | 1 | 1 | 1 | 1 | 1 | DF |
| - | 0 | 1 | 1 | 0 | 0 | 0 | 0 | 0 | 60 | \ | 1 | 1 | 1 | 0 | 0 | 0 | 0 | 0 | E0 |
| / | 0 | 1 | 1 | 0 | 0 | 0 | 0 | 1 | 61 | | 1 | 1 | 1 | 0 | 0 | 0 | 0 | 1 | E1 |
| | 0 | 1 | 1 | 0 | 0 | 0 | 1 | 0 | 62 | S | 1 | 1 | 1 | 0 | 0 | 0 | 1 | 0 | E2 |
| | 0 | 1 | 1 | 0 | 0 | 0 | 1 | 1 | 63 | T | 1 | 1 | 1 | 0 | 0 | 0 | 1 | 1 | E3 |

| Bit: | 0 | 1 | 2 | 3 | 4 | 5 | 6 | 7 | Hex | Bit: | 0 | 1 | 2 | 3 | 4 | 5 | 6 | 7 | Hex |
|---|---|---|---|---|---|---|---|---|---|---|---|---|---|---|---|---|---|---|---|
| | 0 | 1 | 1 | 0 | 0 | 1 | 0 | 0 | 64 | U | 1 | 1 | 1 | 0 | 0 | 1 | 0 | 0 | E4 |
| | 0 | 1 | 1 | 0 | 0 | 1 | 0 | 1 | 65 | V | 1 | 1 | 1 | 0 | 0 | 1 | 0 | 1 | E5 |
| | 0 | 1 | 1 | 0 | 0 | 1 | 1 | 0 | 66 | W | 1 | 1 | 1 | 0 | 0 | 1 | 1 | 0 | E6 |
| | 0 | 1 | 1 | 0 | 0 | 1 | 1 | 1 | 67 | X | 1 | 1 | 1 | 0 | 0 | 1 | 1 | 1 | E7 |
| | 0 | 1 | 1 | 0 | 1 | 0 | 0 | 0 | 68 | Y | 1 | 1 | 1 | 0 | 1 | 0 | 0 | 0 | E8 |
| | 0 | 1 | 1 | 0 | 1 | 0 | 0 | 1 | 69 | Z | 1 | 1 | 1 | 0 | 1 | 0 | 0 | 1 | E9 |
| | 0 | 1 | 1 | 0 | 1 | 0 | 1 | 0 | 6A | | 1 | 1 | 1 | 0 | 1 | 0 | 1 | 0 | EA |
| , | 0 | 1 | 1 | 0 | 1 | 0 | 1 | 1 | 6B | | 1 | 1 | 1 | 0 | 1 | 0 | 1 | 1 | EB |
| % | 0 | 1 | 1 | 0 | 1 | 1 | 0 | 0 | 6C | | 1 | 1 | 1 | 0 | 1 | 1 | 0 | 0 | EC |
| | 0 | 1 | 1 | 0 | 1 | 1 | 0 | 1 | 6D | | 1 | 1 | 1 | 0 | 1 | 1 | 0 | 1 | ED |
| > | 0 | 1 | 1 | 0 | 1 | 1 | 1 | 0 | 6E | | 1 | 1 | 1 | 0 | 1 | 1 | 1 | 0 | EE |
| ? | 0 | 1 | 1 | 0 | 1 | 1 | 1 | 1 | 6F | | 1 | 1 | 1 | 0 | 1 | 1 | 1 | 1 | EF |
| | 0 | 1 | 1 | 1 | 0 | 0 | 0 | 0 | 70 | 0 | 1 | 1 | 1 | 1 | 0 | 0 | 0 | 0 | F0 |
| | 0 | 1 | 1 | 1 | 0 | 0 | 0 | 1 | 71 | 1 | 1 | 1 | 1 | 1 | 0 | 0 | 0 | 1 | F1 |
| | 0 | 1 | 1 | 1 | 0 | 0 | 1 | 0 | 72 | 2 | 1 | 1 | 1 | 1 | 0 | 0 | 1 | 0 | F2 |
| | 0 | 1 | 1 | 1 | 0 | 0 | 1 | 1 | 73 | 3 | 1 | 1 | 1 | 1 | 0 | 0 | 1 | 1 | F3 |
| | 0 | 1 | 1 | 1 | 0 | 1 | 0 | 0 | 74 | 4 | 1 | 1 | 1 | 1 | 0 | 1 | 0 | 0 | F4 |
| | 0 | 1 | 1 | 1 | 0 | 1 | 0 | 1 | 75 | 5 | 1 | 1 | 1 | 1 | 0 | 1 | 0 | 1 | F5 |
| | 0 | 1 | 1 | 1 | 0 | 1 | 1 | 0 | 76 | 6 | 1 | 1 | 1 | 1 | 0 | 1 | 1 | 0 | F6 |
| | 0 | 1 | 1 | 1 | 0 | 1 | 1 | 1 | 77 | 7 | 1 | 1 | 1 | 1 | 0 | 1 | 1 | 1 | F7 |
| | 0 | 1 | 1 | 1 | 1 | 0 | 0 | 0 | 78 | 8 | 1 | 1 | 1 | 1 | 1 | 0 | 0 | 0 | F8 |
| ▲ | 0 | 1 | 1 | 1 | 1 | 0 | 0 | 1 | 79 | 9 | 1 | 1 | 1 | 1 | 1 | 0 | 0 | 1 | F9 |
| : | 0 | 1 | 1 | 1 | 1 | 0 | 1 | 0 | 7A | | 1 | 1 | 1 | 1 | 1 | 0 | 1 | 0 | FA |
| # | 0 | 1 | 1 | 1 | 1 | 0 | 1 | 1 | 7B | | 1 | 1 | 1 | 1 | 1 | 0 | 1 | 1 | FB |
| @ | 0 | 1 | 1 | 1 | 1 | 1 | 0 | 0 | 7C | | 1 | 1 | 1 | 1 | 1 | 1 | 0 | 0 | FC |
| ▲ | 0 | 1 | 1 | 1 | 1 | 1 | 0 | 1 | 7D | | 1 | 1 | 1 | 1 | 1 | 1 | 0 | 1 | FD |
| = | 0 | 1 | 1 | 1 | 1 | 1 | 1 | 0 | 7E | | 1 | 1 | 1 | 1 | 1 | 1 | 1 | 0 | FE |
| " | 0 | 1 | 1 | 1 | 1 | 1 | 1 | 1 | 7F | | 1 | 1 | 1 | 1 | 1 | 1 | 1 | 1 | FF |

DLE = data link escape

DUP = duplicate
EM = end of medium
ENQ = enquiry
EOT = end of transmission
ESC = escape
ETB = end of transmission block
ETX = end of text
EUA = erase unprotected to address
FF = form feed
FM = field mark
IC = insert cursor

ITB = end of intermediate transmission block
NUL = null
PT = program tab
RA = repeat to address
SBA = set buffer address
SF = start field
SOH = start of heading
SP = space
STX = start of text
SUB = substitute
SYN = synchronous
NAK = negative acknowledge

# Index

See "Numbered Devices" index at end of this subject index.